THE
INTERNATIONAL SERIES
OF
MONOGRAPHS ON PHYSICS

SERIES EDITORS

INTERNATIONAL SERIES OF MONOGRAPHS ON PHYSICS

Electronic and Optical Properties of Conjugated Polymers

WILLIAM BARFORD

Department of Physics and Astronomy,
University of Sheffield,
United Kingdom

CLARENDON PRESS · OXFORD
2005

OXFORD

UNIVERSITY PRESS

Great Clarendon Street, Oxford OX2 6DP

Oxford University Press is a department of the University of Oxford.
It furthers the University's objective of excellence in research, scholarship,
and education by publishing worldwide in

Oxford New York

Auckland Cape Town Dar es Salaam Hong Kong Karachi
Kuala Lumpur Madrid Melbourne Mexico City Nairobi
New Delhi Shanghai Taipei Toronto

With offices in

Argentina Austria Brazil Chile Czech Republic France Greece
Guatemala Hungary Italy Japan Poland Portugal Singapore
South Korea Switzerland Thailand Turkey Ukraine Vietnam

Oxford is a registered trade mark of Oxford University Press
in the UK and in certain other countries

Published in the United States
by Oxford University Press Inc., New York

British Library Cataloguing in Publication Data
Data available

Library of Congress Cataloging in Publication Data
Data available

Typeset by the author
Printed in Great Britain
on acid-free paper by
Biddles, King's Lynn

ISBN 0–19–852680–6 (Hbk.)   978–0–19–852680–3

10 9 8 7 6 5 4 3 2 1

PREFACE

Since the discovery of the light emitting properties of the phenyl-based organic semiconductors in 1990 there has been a huge growth of interest in conjugated polymers. The potential device applications are enormous, ranging from optical switching to solar cells and light emitting displays. These new developments are a direct consequence of the active research in the 1980s on conjugated nonelectroluminescent polymers, such as *trans*-polyacetylene. Polyacetylene has particularly interesting and unusual low-lying electronic excitations, and has attracted much experimental and theoretical interest, culminating in the award of the Nobel prize for chemistry in 2000 for research in this field.

The progress in our understanding of the fundamental physics of conjugated polymers, which provides a crucial underpinning to the technological applications, has also been large. This progress has been driven by experimental, theoretical and computational developments. A number of very careful and elegant linear and nonlinear optical spectroscopies over the last decade have established the energies and symmetries of the excited states. Meanwhile, computational advances have been driven by the development of sophisticated numerical techniques, coupled with cheaper and more powerful computers. One of these numerical techniques is the density matrix renormalization group (DMRG) method. This method is highly suited for solving correlated one-dimensional problems.

Conjugated polymers behave as quasi-one dimensional systems owing to their strong intramolecular interactions and rather weak intermolecular interactions. As a consequence, electron-electron interactions are weakly screened, and thus both electron-electron interactions and electron-lattice coupling are fundamentally important in determining the electronic behaviour. Electronic interactions play a crucial role in determining the nature of electronic excitations as they completely change the noninteracting electronic description. Moreover, the coupling of these correlated electronic states to the lattice is also a delicate and complicated problem. Together, electronic interactions and electron-lattice coupling determine the relative energetic ordering of the electronic states, and this, in turn, determines the optical properties of conjugated polymers. This understanding of the origin and nature of the electronic states helps us to explain why some conjugated polymers, for example poly(*para*-phenylene), are electroluminescent, while others, for example *trans*-polyacetylene, are not.

The key aim of this book is to explain how electron-electron interactions and electron-lattice coupling determine the types and character of the low-lying electronic states. Since these effects are complicated, our strategy will be to start with the simplest approximation of noninteracting electrons and gradually develop the full description. At each step care will be taken to explain how electron-electron interactions and electron-lattice coupling modify the predictions of the simpler approximations.

We will see that one of the reasons why understanding the electronic properties of conjugated polymers is such a challenge is because the electronic potential
energy is comparable to the electronic kinetic energy. In other words, the relevant
parameter regime is intermediate between the weak and strong electron-electron
interaction limits. A useful strategy is therefore to tackle these systems from
both the weak and strong coupling extremes. In fact, light emitting polymers
lie on the weak-coupling side of the intermediate regime, whereas nonelectroluminescent polymers (such as *trans*-polyacetylene) lie on the strong coupling
side.

We focus on semiempirical models of π-conjugated systems. There are two
advantages to this strategy over studying *ab initio* models. First, reduced basis models in one dimension can be solved essentially exactly via the DMRG
method for very large systems. Thus, there is no need to make approximations
in the method which might obscure or prejudice an understanding of the physics.
Second, being approximate, reduced basis models retain some symmetries not
present in the *ab initio* models. In particular, particle-hole symmetry is particularly useful in characterizing neutral excited states. Finally, we remark that
although semiempirical, π-electron models are carefully parametrized so that
they also provide accurate predictions of excited state energies.

Solving very large systems by the DMRG method reveals the physics of conjugated polymers not present in conjugated molecules, namely that when the size of
the chain exceeds the spatial extent of the internal structure of the excited states
a quasi-particle description becomes appropriate. Conjugated polymers exhibit
a wealth of different quasi-particles: solitons, excitons, magnon, polarons, etc. It
is an aim of this book to explain the origin and physical consequences of these
quasi-particles.

Nonlinear optical measurements provide the most direct probe of the electronic states. Conversely, the nonlinear susceptibilities can be calculated if there
exists a theoretical understanding of the excited states. We describe the theory
of linear and nonlinear optical processes, and recast the so-called essential states
model in terms of the primary excitons.

Once an understanding of these intramolecular processes is established, another aim of this book will be to explain electronic processes arising from intermolecular interactions. Thus, energy and charge transfer, and excited state
complexes involving two or more polymer chains are described. A mechanism for
determining the singlet exciton yield in light emitting polymers is also discussed.
However, we do not fully address the wider issues of how structure (from the local
polymer packing to the global morphology) affects the performance of systems
comprising conjugated polymers. Another important consequence of interchain
excitations is that they significantly modify the energy of some intrachain excitations. An understanding of this effect is crucial to the interpretation of optical
experiments.

A final aim of this book is to demonstrate how our theoretical understanding of excited states enables us to make a consistent interpretation of experimental results. Two final chapters draw these themes together in discussing *trans*-polyacetylene, and the hugely technologically important phenyl-based light emitting polymers.

The book is therefore organized as follows. Chapter 1 gives a brief overview of the electronic properties of conjugated polymers. Our basic models for describing these properties are semiempirical π-electron models. So, Chapter 2 introduces and motivates these models. Next, we consider the solution of these models in various limits: noninteracting electrons with fixed geometry in Chapter 3, noninteracting electrons with electron-lattice coupling in Chapter 4, and interacting electrons with fixed geometry in Chapter 5. Chapter 6 is devoted to a discussion of excitons, as these are so important in determining the photophysical properties of conjugated polymers. The electronic states of interacting electrons with electron-lattice coupling are described Chapter 7. Chapters 8 and 9 describe optical and electronic processes in conjugated polymers, respectively, while experimental and theoretical investigations of *trans*-polyacetylene and light emitting polymers are described in Chapters 10 and 11. Chapter 8 introduces the nonlinear optical spectroscopies that are used to identify the excited states of conjugated polymers discussed in Chapters 10 and 11.

This book was written with two kinds of readers in mind. One kind of readers are experimentalists who wish to understand and interpret their experimental data in terms of the fundamental electronic and optical properties of conjugated polymers. The other type of readers are theoretical and computational chemists and physicists who both want to understand the fundamental properties of conjugated polymers, as well as wishing to develop models and perform calculations of their own. For these readers there are a number of appendices containing material too technical for the main chapters. In particular, Appendix H gives a brief review of the DMRG method.

Sheffield   William Barford
January 2005

ACKNOWLEDGEMENTS

My interest in conjugated polymers began in the 1995 when Donal Bradley established a polymer optoelectronics research group in the Department of Physics and Astronomy at the University of Sheffield. At the same time three theorists then at Sheffield, Robert Bursill, Gillian Gehring and Tao Xiang, were using Steve White's highly-accurate density matrix renormalization group (DMRG) method to solve one-dimensional correlated electron problems. The opportunity to apply the DMRG method to models of conjugated polymers at the same time as developing a collaboration with the experimental group was too great to miss.

Since then I have learnt a great deal about conjugated polymers from a large number of people. Donal Bradley, Paul Lane (both then in Sheffield) and Simon Martin are three experimentalists who patiently answered many questions on the optoelectronic properties of conjugated polymers.

Collaborations and/or conversations with David Beljonne, Peter Bobbert, Richard Friend, Neil Greenham, Anna Köhler, Mikhail Yu Lavrentiev, David Lidzey, Sumit Mazumdar, Carlos Silva, Markus Wohlgenannt, and David Yaron are all greatly appreciated.

Last, but not least, my sincere thanks to my coworker since 1995, Robert Bursill. It was he who principally developed the DMRG programs that we have used to study correlated models of conjugated polymers. This book was inspired by those studies.

This book was started in May, 2002. Without the benefit of a research fellowship from the Leverhulme Trust in 2003-04 it is unlikely that it would ever have been completed. I am very grateful to the Leverhulme Trust for the opportunity to be relieved from my teaching duties for one year. I would also like to thank the Engineering and Physical Sciences Research Council, the Gordon Godfrey Bequest of the University of New South Wales and the Royal Society for sponsoring my research in this subject. Thanks too to the University of New South Wales, and the Cavendish Laboratory and Clare Hall, Cambridge for their hospitality in 2003-04.

I would like to thank Sonke Adlung at Oxford University Press for his enthusiasm for the book proposal, and his colleagues at OUP and Mark Fox in Sheffield for their help over the preparation of the manuscript. Finally, I am very grateful to Simon Martin and Sumit Mazumdar for providing valuable critical feedback on the draft manuscript. As always, however, all errors and omissions are the sole responsibility of the author.

This book is dedicated to my mother and father.

CONTENTS

xi

1

INTRODUCTION TO CONJUGATED POLYMERS

Research into the electronic, optical, and magnetic properties of conjugated polymers began in the 1970s after a number of seminal experimental achievements. First, the synthesis of polyacetylene thin films (Itô *et al.* 1974) and the subsequent success in doping these polymers to create conducting polymers (Chiang *et al.* 1977) established the field of synthetic metals. Second, the synthesis of the phenyl-based polymers and the discovery of electroluminescence under low voltages in these systems (Burroughes *et al.* 1990) established the field of polymer optoelectronics.

The electronic and optical properties of conjugated polymers, coupled with their mechanical properties and intrinsic processing advantages, means that they are particularly attractive materials for the electronics industry. There are many potential applications including, light emitting devices, nonlinear optical devices, photovoltaic devices, plastic field-effect transistors, and electro-magnetic shielding. The discovery and development of conductive polymers was recognized by the award of the Nobel prize for chemistry in 2000 to Heeger, MacDiarmid, and Shirakawa (see Heeger 2000; MacDiarmid 2000; and Shirakawa 2000).

A conjugated polymer is a carbon-based macromolecule through which the valence π-electrons are delocalized.[1] *Trans*-polyacetylene, illustrated in Fig. 1.1, is a linear polyene, whose ground state structure is composed of alternating long and short bonds. Also shown in Fig. 1.1 are two other linear polyenes, *cis*-polyacetylene and polydiacetylene. The light emitting polymers, for example, poly(*para*-phenylene) (or PPP) and poly(*para*-phenylene vinylene) (or PPV), are characterized by containing a phenyl ring in their repeat units. PPP and PPV are illustrated in Fig. 1.2.

As well as their many important technological applications, conjugated polymers are also active components in many biological optophysical processes, for example, as light collectors in photosynthesis, and in the vision mechanism via photoisomerization. Charge transport in organic molecules is also an important component of cellular function. Thus, many of the concepts developed in this book are applicable to these biological systems.

Conjugated polymers exhibit electronic properties that are quite different from those observed in the corresponding inorganic metals or semiconductors. These unusual electronic properties may essentially be attributed to fact that conjugated polymers behave as quasi-one dimensional systems owing to their

---

[1] Conjugate from the Latin *conjugatus*, meaning to join or unite.

trans-polyacetylene

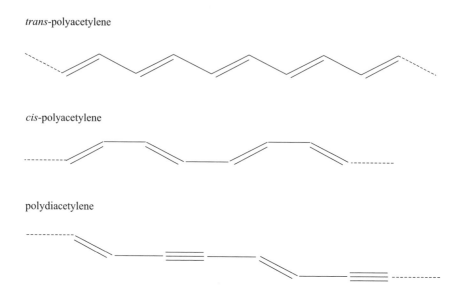

cis-polyacetylene

polydiacetylene

FIG. 1.1. The carbon backbone of some linear polyenes. The hydrogen atoms are not shown. More detailed chemical structures are illustrated in Chapter 2.

strong intramolecular electronic interactions and relatively weak intermolecular electronic interactions. Weak intermolecular electronic interactions (arising from poor electronic wavefunction overlap) coupled to strong dissipation (or dephasing) mechanisms means that quantum mechanical coherence is generally confined to a single chain, or at most a few other chains. Electronic wavefunctions are therefore typically localized on single chains, or to pairs of chains in the case of excited state complexes. This quasi-one dimensionality also means that electron-electron interactions are weakly screened. Thus, electronic correlations are important in determining the character of the electronic states. A final important factor in determining the character of the electronic states is that the electrons and lattice are strongly coupled. As for electron-electron interactions, the effects of electron-lattice coupling are enhanced in low dimensions.

Much early theoretical work on conjugated oligomers and polymers treated electron-electron and electron-lattice interactions independently. In the 1950s the focus was on the role of electron-lattice interactions in causing a metal-semiconductor transition in one-dimensional metals (Fröhlich 1954; Peierls 1955), and in determining the bond alternation in linear polyenes (Ooshika 1957, 1959; Longuet-Higgins and Salem 1959). It was also realized that a broken-symmetry ground state of bond alternation implies bond defects between different domains of bond alternation (Ooshika 1957; Longuet-Higgins and Salem 1959), and to associated mid-gap electronic states (Pople and Walmsley 1962). Theoretical and experimental investigations into excited states and their associated bond

poly(*para*-phenylene)

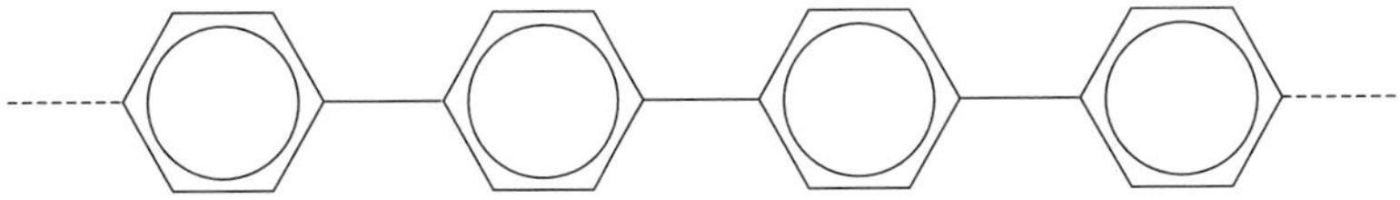

poly(*para*-phenylene-vinylene)

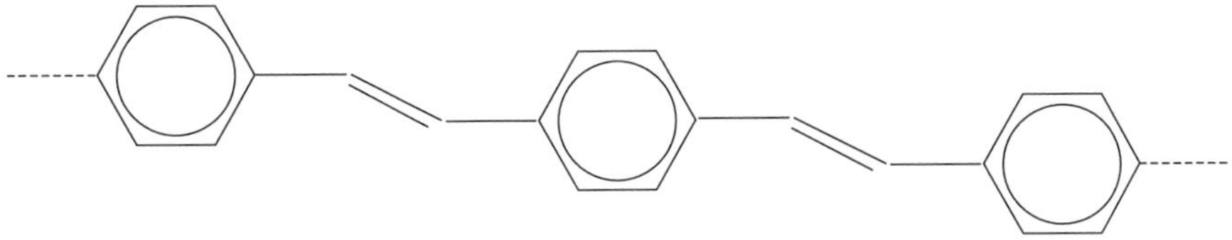

FIG. 1.2. The carbon backbone of some phenyl-based light emitting polymers.

defects (or solitons) grew rapidly after the introduction of a simplified model of electron-lattice interactions in *trans*-polyacetylene by Su, Schrieffer, and Heeger (Su *et al.* 1979). These developments are reviewed in (Heeger *et al.* 1988).

An alternative point of view, namely that electron-electron interactions are important in determining the electronic properties in conjugated polymers, was advocated by Ovchinnikov and coworkers (Ovchinnikov *et al.* 1973), who argued that electronic correlations are principally responsible for the optical gap in linear polyenes. Likewise, the lack of electroluminescence in linear polyenes was attributed to a dipole-forbidden strongly correlated singlet state lying below the dipole-allowed singlet (Hudson and Kohler 1972; Schulten and Karplus 1972). Another indication of the importance of electron-electron interactions is that electroluminescence from light emitting polymers occurs from exciton (or bound particle-hole states), and not from a direct band to band transition.

In fact, as we emphasize in this book, both electron-electron and electron-lattice interactions must be treated together in order to achieve a coherent description of the excited states of conjugated polymers. It is the interplay of these two processes that leads to the rich variety and relative energetic ordering of the electronic states in conjugated polymers, and ultimately to their electronic and optical properties. For example, strong electronic interactions *and* electron-lattice coupling in *trans*-polyacetylene reverses the energy of the dipole allowed and forbidden singlet states, rendering it nonelectroluminescent.

As a consequence of their size, conjugated polymers exhibit some physical properties more associated with solid state physics than molecular physics. One of these concepts is broken symmetry ground states. Another is the quasi-particle description of excited states, widely used in solid physics, which is also appropriate in conjugated polymers provided that the conjugation length is longer than the internal spatial extent of the excited state. Since the conjugation length is typically $10 - 20$ repeat units (or $\sim 15 - 50$ Å) (much longer than a conjugated molecule) while the typical size of the internal structure of an excited state is $5 - 10$ Å, conjugated polymers do indeed exhibit a wealth of quasi-particles, including solitons, excitons, magnons, and polarons. It is an aim of this book to explain the origin and physical consequences of these quasi-particles.

Undoped (neutral) conjugated polymers are semiconductors, with optical gaps of $\sim 2 - 3$ eV and charge (or band) gaps typically $\sim 0.5 - 1.0$ eV higher in energy, reflecting the large exciton binding energies in polymers. Doped polymers have 'metallic' conductivities of typically $10^3 - 10^5$ S cm^{-1}, with the highest conductivity in *trans*-polyacetylene being 10^5 S cm^{-1}. (In comparison, the conductivity of room temperature copper is 10^6 S cm^{-1}.) However, although the conductivities of conjugated polymers are reasonable, their performance as synthetic metals is adversely affected both by disorder (which means that the conductivity is close to the localization transition), and by the unstable nature of highly doped polymers (Heeger 2000).

This book is principally concerned with neutral, semiconducting conjugated polymers. Generally, as already remarked, these have quite different properties from their inorganic counterparts. Inorganic semiconductors are characterized by strong electron-electron screening ($\epsilon \sim 11$ in silicon) and strong spin-orbit coupling. Since (in the effective-mass exciton model) exciton binding energies $\sim \epsilon^{-2}$ and electron-hole separations $\sim \epsilon$, excitons in inorganic semiconductors are weakly bound (~ 25 meV) with large particle-hole separations (~ 50 Å). The strong spin-orbit coupling in inorganic semiconductors means that the total angular momentum is a good quantum number, and thus that singlet and triplet states are mixed. In contrast, electron-electron screening is weak in organic systems ($\epsilon \sim 2 - 3$), so correspondingly the exciton binding energies are large ($\sim 0.5 - 1.5$ eV) and the particle-hole separations are small ($\sim 5 - 10$ Å).

Spin-orbit coupling in organic systems is small, and thus spin is a good quantum number. The large exchange energy between the lowest-lying singlet and triplet excitons is another consequence of strong electron-electron interactions in conjugated polymers. Definite spin states also have an important implication in light emitting devices, as it implies that if electron-hole recombination is spin-independent then only one-quarter of the injected electron-hole pairs become singlet excitons. The maximum electroluminescence efficiency would therefore be 25%. However, while spin-independent recombination does appear to occur in conjugated molecules, it is a matter of current controversy in conjugated polymers. An understanding of the excited state spectrum and recombination processes is a prerequisite to address this issue.

A topic that has bedevilled the field is the exciton binding energy. There are a number of reasons why this subject is controversial. First, to a certain extent the concept is ill-defined. The concept of an exciton binding energy only makes sense when the energy of a bound electron-hole quasi-particle can be compared to a widely-separated uncorrelated electron-hole pair. However, if the electron-hole separation of the 'bound' state is of the order of the oligomer size it is not possible to distinguish between that energy and the energy of a widely separated pair. Second, there is a difference between vertical and relaxed energies. Thus, the energy difference between the vertical transition to the exciton and the vertical transition to an unbound electron and hole pair is different from the energy difference between the relaxed exciton-polaron and a pair of polarons. Since the exciton relaxation energy is roughly the same as that of a single polaron (namely $\sim 0.1-0.2$ eV), this gives a difference of binding energies of $\sim 0.1-0.2$ eV between these two estimates. Finally, there is a significant difference (of roughly $1-2$ eV) between the exciton binding energy in an isolated polymer chain and the binding energy in a polymer chain in the solid state. This is because dielectric screening of the intramolecular excited states by the environment significantly modifies their energies. However, in contrast to inorganic semiconductors, since the binding energy of the lowest energy excitons is comparable to their excitation energies, this screening involves both dispersion and solvation components. It therefore cannot simply be modelled by a static screened electron-hole potential, and it is thus more difficult to theoretically predict the bulk binding energy. In this book we take the view that spectroscopic probes are a reliable means of distinguishing excited states. Thus, equipped with a theoretical understanding of the excited states we use these probes to estimate the exciton binding energies.

The electronic and optical properties of conjugated polymers briefly described in this chapter are summarized in Table 1.1. It is these properties that are the subject of this book. We investigate semiempirical models of π-conjugated systems to address the following issues:

- What are the roles of electron-electron and electron-lattice interactions in determining the broken symmetry ground states, and the type, character and relative energetic ordering of the excited states?
- How does the type and character of the electronic states determine the electronic and optical processes in conjugated polymers? For example,
  * How do they determine the nonlinear optical spectroscopies?
  * How do they determine energy and charge transport?
- How can nonlinear optical spectroscopies be used to determine the character of the excited states?
- What is the role of the polymer's environment in modifying the energy and character of the excited states?

We begin these investigations by describing π-electron theories of conjugated polymers in the next chapter.

Table 1.1 *Summary of the electronic and optical properties of conjugated polymers*

Physics	Consequences	Chapter(s)
Quasi-one dimensional systems.	The effects of electron-electron and electron-lattice interactions are enhanced. Broken symmetry ground states.	4, 5, 7
Conjugation lengths $> 10 - 20$ repeat units.	Quasi-particle description of excited and charged states, for example, solitons, excitons, magnons, and polarons.	$4 - 11$
Weak electron-electron screening (the dielectric constant in the solid state is $\epsilon \sim 2 - 3$).	The lowest energy excitons have large binding energies ($\sim 0.5 - 1.5$ eV) and small particle-hole separations ($\sim 5 - 10$ $\mathring{A}$).	6, 10, 11
Small exciton sizes.	Excitons are generally localized on single chains and luminesce strongly if they are the lowest excited singlet excitation.	8
Strong electron-lattice coupling.	Solitonic and polaronic structures. Self-trapping of excited states.	4, 7, 10, 11
Electron-lattice interactions coupled to strong electron-electron interactions.	Enhanced bond alternation, energy reversal of excited states and four-soliton excited states in linear polyenes.	7, 10
Energy reversal of excited states in linear polyenes.	Not electroluminescent.	7, 10
Weak spin-orbit coupling and strong electron-electron interactions.	Spin is a good quantum number. Large triplet-singlet exchange energies.	6, 10, 11
Weak interchain electronic coupling.	Interchain excited state complexes (e.g. excimers and exciplexes), which may quench luminescence. Dephasing of intermolecular wavefunctions.	9
Exciton binding energies comparable to the optical gap.	Dielectric screening of excitons from both solvation and dispersion interactions.	9
Disorder.	Doped conjugated polymers exhibit conductivity close to a disorder-induced metal-insulator transition. Energy spectra are broadened.	

2

π-ELECTRON THEORIES OF CONJUGATED POLYMERS

2.1 Introduction

The aim of this chapter is to introduce some of the language and notation used throughout this book. We begin by discussing the many body Hamiltonian that describes the electronic and nuclear degrees of freedoms in polymers. Then we introduce the Born-Oppenheimer approximation which treats the nuclear degrees of freedom adiabatically and leads to the Born-Oppenheimer Hamiltonian for the electronic degrees of freedom. Next, we introduce the concept of $sp^{(n)}$ hybridization that decouples the higher energy σ electronic processes from the lower energy π electronic processes. This leads to the concept of π-conjugation and allows us to introduce the π-electron models of conjugated polymers. These models include the Hückel model of noninteracting electrons, the Su-Schrieffer-Heeger model of electron-phonon coupling, and models of interacting electrons, such as the Pariser-Parr-Pople model. Finally, we discuss the various symmetries, particularly spatial and particle-hole symmetries, that characterize the electronic states of conjugated polymers.

2.2 The many body Hamiltonian

The electronic and nuclear degrees of freedom of a system are described by the many body Hamiltonian,

$$H_T = H_{n-n}(\{\mathbf{R}\}) + H_{e-e}(\{\mathbf{r}\}) + H_{e-n}(\{\mathbf{r}\}, \{\mathbf{R}\}), \tag{2.1}$$

where

$$H_{n-n}(\{\mathbf{R}\}) = \sum_\alpha \frac{P_\alpha^2}{2M_\alpha} + \frac{1}{2} \sum_{\alpha \neq \beta} \frac{Z_\alpha Z_\beta e^2}{|\mathbf{R}_\alpha - \mathbf{R}_\beta|} \tag{2.2}$$

describes the kinetic energy of the nuclei and their mutual potential energy from Coulomb interactions,

$$H_{e-e}(\{\mathbf{r}\}) = \sum_i \frac{p_i^2}{2m_i} + \frac{1}{2} \sum_{i \neq j} \frac{e^2}{|\mathbf{r}_i - \mathbf{r}_j|} \tag{2.3}$$

describes the kinetic energy of the electrons and their mutual potential energy, and

$$H_{e-n}(\{\mathbf{r}\}, \{\mathbf{R}\}) = -\frac{1}{2} \sum_{\alpha,i} \frac{Z_\alpha e^2}{|\mathbf{R}_\alpha - \mathbf{r}_i|} \tag{2.4}$$

describes the potential energy arising from the Coulomb interactions between the nuclei and electrons. $\{\mathbf{R}\}$ and $\{\mathbf{r}\}$ represent the set of nuclear and electronic coordinates, respectively, while M and m are the nuclear and electronic masses, respectively. Z_α is the nuclear number of the αth nucleus and e is the electronic charge.

The full Hamiltonian, H_T, can only be solved exactly for the hydrogen atom in free space. For all other more complex systems various approximation schemes are required. A very important approximation scheme is the Born-Oppenheimer approximation, whereby the electronic degrees of freedom are explicitly decoupled from the nuclear dynamics. This will be discussed in the next section. Another approximation scheme for conjugated polymers is to focus explicitly on the low-energy electronic degrees of freedom using parameters that in principle are determined by the high-energy electronic degrees of freedom, but in practice are treated semiempirically. This leads to effective π-electron models, which often accurately describe the electronic states of conjugated systems. The most useful consequence of π-electron models, however, is that they provide a quantitative description of the low energy physics.

2.3   The Born-Oppenheimer approximation

The Born-Oppenheimer approximation exploits the fact that the nuclear mass is very much larger than the electronic mass, and therefore the nuclear dynamics are expected to be 'slow' in comparison to the electronic dynamics. Thus, it is convenient to introduce an electronic state, $|i; \{\mathbf{R}\}\rangle$, that is determined by a set of *static* nuclear coordinates, $\{\mathbf{R}\}$. $|i; \{\mathbf{R}\}\rangle$ thus depends *parametrically* on $\{\mathbf{R}\}$.

The Born-Oppenheimer approximation is to assume that a many body state, $|I\rangle$, may be factorized as a *single*, direct product of an electronic state, $|i; \{\mathbf{R}\}\rangle$, and a nuclear state $|\nu_i\rangle$ associated with the electronic state, $|i; \{\mathbf{R}\}\rangle$:

$$|I\rangle = |i; \{\mathbf{R}\}\rangle|\nu_i\rangle. \tag{2.5}$$

Then $|I\rangle$ satisfies an eigenvalue equation and $|i; \{\mathbf{R}\}\rangle$ is an eigenstate of the so-called Born-Oppenheimer Hamiltonian, provided that that the electronic state $|i; \{\mathbf{R}\}\rangle$ is so weakly parametrized by the nuclear coordinates that,

$$P_\alpha|i; \{\mathbf{R}\}\rangle \approx P_\alpha^2|i; \{\mathbf{R}\}\rangle \approx 0. \tag{2.6}$$

To show this, consider the action of the full Hamiltonian, H_T, on $|I\rangle$:

$$H_T|I\rangle = \left(H_{e-e}(\{\mathbf{r}\}) + H_{e-n}(\{\mathbf{r}\}, \{\mathbf{R}\}) + \frac{1}{2}\sum_{\alpha \neq \beta}\frac{Z_\alpha Z_\beta e^2}{|\mathbf{R}_\alpha - \mathbf{R}_\beta|} \right)|i; \{\mathbf{R}\}\rangle|\nu_i\rangle$$
$$+ \sum_\alpha \frac{P_\alpha^2}{2M_\alpha}|i; \{\mathbf{R}\}\rangle|\nu_i\rangle, \tag{2.7}$$

where we have explicitly decomposed $H_{n-n}(\{\mathbf{R}\})$ into its separate kinetic and potential energy components. Now, using the product rule and the approximations encapsulated by eqn (2.6),

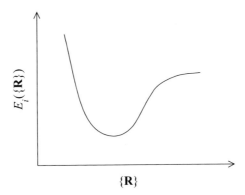

FIG. 2.1. The adiabatic potential energy surface, $E_i(\{\mathbf{R}\})$.

$$P_\alpha^2 \left(|i;\{\mathbf{R}\}\rangle|\nu_i\rangle\right) = |\nu_i\rangle P_\alpha^2|i;\{\mathbf{R}\}\rangle + |i;\{\mathbf{R}\}\rangle P_\alpha^2|\nu_i\rangle + 2P_\alpha|i;\{\mathbf{R}\}\rangle P_\alpha|\nu_i\rangle$$
$$\approx |i;\{\mathbf{R}\}\rangle P_\alpha^2|\nu_i\rangle. \tag{2.8}$$

Thus, eqn (2.7) becomes,

$$H_T|I\rangle \approx H_{\mathrm{BO}}(\{\mathbf{r}\};\{\mathbf{R}\})|i;\{\mathbf{R}\}\rangle|\nu_i\rangle + |i;\{\mathbf{R}\}\rangle \sum_\alpha \frac{P_\alpha^2}{2M_\alpha}|\nu_i\rangle, \tag{2.9}$$

where we define the Born-Oppenheimer Hamiltonian, H_{BO}, as,

$$H_{\mathrm{BO}}(\{\mathbf{r}\};\{\mathbf{R}\}) = H_{e-e} + H_{e-n} + \frac{1}{2}\sum_{\alpha\neq\beta}\frac{Z_\alpha Z_\beta \mathrm{e}^2}{|\mathrm{R}_\alpha - \mathrm{R}_\beta|}. \tag{2.10}$$

By definition, $|i;\{\mathbf{R}\}\rangle$ is an eigenstate of H_{BO} whose corresponding eigenvalue, $E_i(\{\mathbf{R}\})$, is the sum of the electronic kinetic energy and all the potential energy terms. $E_i(\{\mathbf{R}\})$ is also known as the *adiabatic potential energy surface*, and is shown schematically in Fig. 2.1. As we shall see shortly, $E_i(\{\mathbf{R}\})$ is the *effective potential* experienced by the nuclei.

Finally, inserting

$$H_{\mathrm{BO}}|i;\{\mathbf{R}\}\rangle = E_i(\{\mathbf{R}\})|i;\{\mathbf{R}\}\rangle \tag{2.11}$$

into eqn (2.9) we have,

$$H_T|I\rangle \approx |i;\{\mathbf{R}\}\rangle\left(E_i(\{\mathbf{R}\} + \sum_\alpha \frac{P_\alpha^2}{2M_\alpha})\right)|\nu_i\rangle = \epsilon|i;\{\mathbf{R}\}\rangle|\nu_i\rangle. \tag{2.12}$$

ϵ is the sum of the nuclear kinetic energy and the effective potential experienced by the nuclei namely, $E_i(\{\mathbf{R}\})$.

The Born-Oppenheimer approximation is an adiabatic approximation, as it is equivalent to the assumption that there are no transitions between the electronic states from changes of the nuclear coordinates. Mixing of the Born-Oppenheimer states, $\{|I\rangle\}$, occurs by the nonadiabatic term neglected in eqn (2.9), namely,

$$\frac{1}{2M_\alpha} \sum_\alpha \left(2P_\alpha|i; \{\mathbf{R}\}\rangle P_\alpha|\nu_i\rangle + |\nu_i\rangle P_\alpha^2|i; \{\mathbf{R}\}\rangle \right). \tag{2.13}$$

These nonadiabatic processes have important consequences, such as interconversion and hence energy relaxation processes, as well as in chemical reactions.

The Born-Oppenheimer Hamiltonian is widely used in solid state physics and quantum chemistry to study the electronic properties of materials - and it is also widely used in this book. In the next section we recast it in the very convenient second quantization representation.

2.4   Second quantization of the Born-Oppenheimer Hamiltonian

The Born-Oppenheimer Hamiltonian describes the electronic degrees of freedom. A convenient representation of fermion Hamiltonians is by second quantization. As this representation is widely used in this book, we give a brief discussion of it here. A good discussion may be found in (Landau and Lifshitz 1977) or (Surján 1989).

In Dirac notation we may represent a single-particle electronic state as the ket $|i\rangle$. Suppose that the single-particle states $\{|i\rangle\}$ form an orthonormal basis. The projection of $|i\rangle$ onto the coordinate representation, $\{|\mathbf{r}\rangle\}$, (where $|\mathbf{r}\rangle$ is an eigenstate of the position operator, $\hat{\mathbf{r}}$) gives the single-particle wave function (or orbital), $\phi_i(\mathbf{r})$, namely

$$\phi_i(\mathbf{r}) \equiv \langle \mathbf{r}|i\rangle. \tag{2.14}$$

It is often convenient to regard $|i\rangle$ and $\phi_i(\mathbf{r})$ as different, but essentially equivalent representations of a single-particle state.

Then we may define the *creation operator*, $c_i^\dagger$, such that it creates an electron in the orbital $\phi_i(\mathbf{r})$. Formally,

$$|i\rangle = c_i^\dagger|0\rangle, \tag{2.15}$$

where $|0\rangle$ is the vacuum state. The adjoint to the creation operator, the *annihilation operator*, c_i, destroys the electron in $\phi_i(\mathbf{r})$.

Since electrons carry spin we also need to define the creation operator $c_{i\sigma}^\dagger$, which creates an electron with spin σ in the spin-orbital,

$$\chi_i(\mathbf{r}, \sigma) = \phi_i(\mathbf{r})\sigma. \tag{2.16}$$

σ is the two-component spinor with values of $\binom{1}{0}$ for an up-spin and $\binom{0}{1}$ for a down-spin with respect to an arbitrary axis of quantization. Similarly, $c_{i\sigma}$ destroys an electron with spin σ in $\chi_i(\mathbf{r}, \sigma)$.

The *number operator*, $N_{i\sigma} = c_{i\sigma}^\dagger c_{i\sigma}$, counts the number of electrons with spin σ in $\chi_i(\mathbf{r}, \sigma)$.

The *Pauli principle*, that the many body fermion wavefunction must be anti-symmetric with respect to an exchange of coordinates, implies that the creation and annihilation operators satisfy the following anticommutation relations:

$$c_{i\sigma}c_{j\sigma'}^{\dagger} + c_{j\sigma'}^{\dagger}c_{i\sigma} = \delta_{ij}\delta_{\sigma\sigma'} \tag{2.17}$$

and

$$c_{i\sigma}c_{j\sigma'} + c_{j\sigma'}c_{i\sigma} = c_{i\sigma}^{\dagger}c_{j\sigma'}^{\dagger} + c_{j\sigma'}^{\dagger}c_{i\sigma}^{\dagger} = 0. \tag{2.18}$$

Using these rules, it can be shown that in second quantization the Born-Oppenheimer Hamiltonian is expressed as,

$$H_{\text{BO}} = \sum_{ij\sigma} \tilde{t}_{ij}\left(c_{i\sigma}^{\dagger}c_{j\sigma} + c_{j\sigma}^{\dagger}c_{i\sigma}\right) + \frac{1}{2}\sum_{ijkl\sigma\sigma'} \tilde{V}_{ijkl}c_{i\sigma}^{\dagger}c_{k\sigma'}^{\dagger}c_{l\sigma'}c_{j\sigma}$$

$$+\frac{1}{2}\sum_{\alpha\neq\beta}\frac{Z_{\alpha}Z_{\beta}e^2}{|\mathbf{R}_{\alpha} - \mathbf{R}_{\beta}|}, \tag{2.19}$$

where

$$\tilde{t}_{ij} = \int \phi_i^*(\mathbf{r})\left[\frac{\mathbf{p}^2}{2m} - \sum_{\alpha}\frac{Z_{\alpha}e^2}{|\mathbf{R}_{\alpha} - \mathbf{r}|}\right]\phi_j(\mathbf{r})d^3\mathbf{r}, \tag{2.20}$$

is the one-electron integral, and

$$\tilde{V}_{ijkl} = \int\int \phi_i^*(\mathbf{r})\phi_k^*(\mathbf{r}')\frac{e^2}{|\mathbf{r} - \mathbf{r}'|}\phi_l(\mathbf{r}')\phi_j(\mathbf{r})d^3\mathbf{r}d^3\mathbf{r}', \tag{2.21}$$

is the two-electron integral.

We may interpret the terms in eqn (2.19) as follows. The first term on the right-hand side represents the transfer of an electron from the spin-orbital $\chi_j(\mathbf{r}, \sigma)$ to the spin-orbital $\chi_i(\mathbf{r}, \sigma)$ (and vice versa), with an energy scale t_{ij}. The terms $i = j$ in the sum represent the single-particle on-site energy,[2] while the other terms represent the hybridization of the electrons between different orbitals.[3] The second term on the right-hand side represents electron-electron interactions, the most important being the direct Coulomb interaction when $i = j$ and $k = l$, as we discuss in Section 2.6. For readers not familiar with the second quantization approach, Appendix A describes a first quantization representation of the first term on the right-hand side of eqn (2.19).

Equation (2.19) is a formal representation of all the electronic degrees of freedom. It is necessary, and indeed useful when considering low energy processes, to truncate the basis. For carbon-based systems this may be conveniently accomplished by sp^n hybridization, as we describe in the next section.

2.5 sp^n hybridization

Atomic orbital hybridization is a well-known approximate procedure in quantum chemistry designed to understand the nature of chemical bonds. In this procedure

---

[2]Denoted by α in quantum chemistry text books.

[3]Called the resonance integral, and denoted by β in quantum chemistry text books.

(a)

$$H \text{——} C \equiv C \text{——} H$$

(b)

$$\begin{array}{c} H \\ \diagdown \\ H \diagup \end{array} C = C \begin{array}{c} \diagup H \\ \diagdown \\ H \end{array}$$

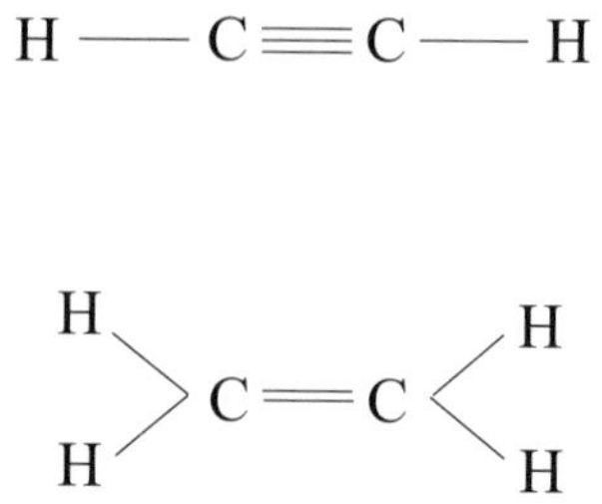

FIG. 2.2. The chemical structures of acetylene (a) and ethylene (b).

linear combinations of atomic orbitals are constructed that have a directionality optimized for bonding. In carbon-based molecules the orbitals involved in the hybridization are the four outer valence orbitals, namely the $2s$, $2p_x$, $2p_y$, and $2p_z$ orbitals. There are three types of sp^n hybridization: sp hybridization, found in linear molecules such as acetylene (shown in Fig. 2.2(a)); sp^2 hybridization, found in planar molecules such as ethylene (shown in Fig. 2.2(b)); and sp^3 hybridization, found in three-dimensional structures, such as methane. We briefly describe these below. For further details, see (Coulson 1961), (Cohen-Tannoudji *et al.* 1977), or (Atkins and Friedman 1997).

2.5.1 *sp hybridization*

We start with a discussion of sp hybridization. Consider a bond between two carbon atoms oriented along the x-axis, as shown in Fig. 2.3. Then, the two sp hybrids per carbon atom, $|\sigma_{\pm}\rangle$, are formed from the $2s$ and $2p_x$ orbitals as follows:

$$|\sigma_{\pm}\rangle = \frac{1}{\sqrt{2}} \left(|2s\rangle \pm |2p_x\rangle \right). \tag{2.22}$$

The σ orbitals are highly directional, and result in the strong covalent bonding. The remaining orbitals, $2p_y$ and $2p_z$, remain unhybridized and are known as π orbitals. They are orthogonal to the σ orbitals, and hinder rotations around the bond axis. Figure 2.3 shows the acteylene structure with the pair of σ orbitals and two pairs of π orbitals between the pair of carbon atoms, which altogether form a 'triple' bond.

2.5.2 *sp² hybridization*

In sp^2 hybridization there are three sp^2 hybrids per carbon atom. $|\sigma_1\rangle$, illustrated in Fig. 2.4, is constructed from the $2s$ and $2p_x$ orbitals as follows:

$$|\sigma_1\rangle = \frac{1}{\sqrt{3}} \left(|2s\rangle + \sqrt{2}|2p_x\rangle \right). \tag{2.23}$$

The remaining σ orbitals are constructed from the $2s$, $2p_x$, and $2p_y$, orbitals, and may be defined as:

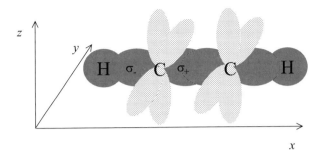

FIG. 2.3. A schematic illustration of the σ and π ($2p_y$ and $2p_z$) orbitals (shown hatched) in acetylene.

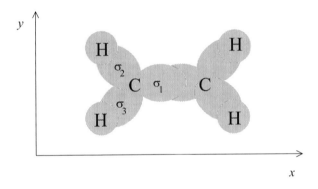

FIG. 2.4. A schematic illustration of the σ orbitals in ethylene. The π ($2p_z$) orbitals (not shown) are normal to the paper.

$$|\sigma_2\rangle = \exp\left(-i\frac{2\pi}{3}\frac{L_z}{\hbar}\right)|\sigma_1\rangle \tag{2.24}$$

and

$$|\sigma_3\rangle = \exp\left(i\frac{2\pi}{3}\frac{L_z}{\hbar}\right)|\sigma_1\rangle, \tag{2.25}$$

where L_z is the z-component of the angular moment operator. Thus, by definition, each σ orbital is oriented $\pm120^0$ with respect to each other. The remaining $2p_z$ (or π) orbital lies perpendicular to the plane, and hinders rotations around the carbon-carbon bond axis. The pairs of σ and π orbitals between each pair of carbon atoms form a 'double' bond.

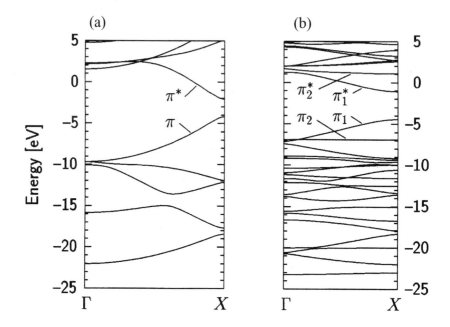

FIG. 2.5. *Ab initio* calculated quasi-particle band structure. (a) *trans*-polyacetylene, where the valence and conduction bands are denoted as π and π^*, respectively, and the four bands below the valence band are formed from the three sp^2 hybrids and the hydrogen $1s$ orbital. (b) poly(*para*-phenylene vinylene), where the valence, π_1, conduction, π_1^*, and nonbonding bands, π_2 and π_2^*, are shown. Reprinted with permission from M. Rohlfing and S. G. Louie, *Phys. Rev. Lett.* **82**, 1959, 1999. Copyright 1999 by the American Physical Society.

2.5.3 sp^3 hybridization

Here all four valence orbitals are involved in sp^3 hybridization, resulting in a tetrahedral orientation of the bonds, as found in methane or ethane. There is one pair of σ orbitals between each pair of carbon atoms, resulting in a 'single' bond.

2.5.4 *Remarks*

The electrons in σ orbitals are localized in the σ bonds, whereas it is possible for electrons in the π orbitals to delocalize (or *conjugate*) throughout the molecule. Molecules composed of sp or sp^2 hybridized orbitals are therefore known as '*unsaturated*', while molecules composed of sp^3 hybrids are known as '*saturated*'. Conjugated polymers are typically sp^2 hybridized with one π-orbital per CH group. (An exception is polydiacetylene which has both sp and sp^2 hybridization.) The four valence electrons are shared amongst the four hybrid orbitals,

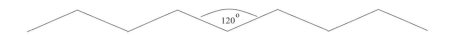

FIG. 2.6. A schematic representation of a conjugated polymer with sp^2 hybridization as described by π-electron models. Each vertex is a site with on average one π-electron.

and thus there is one π-electron per CH group.

The σ and π orbitals are formally decoupled at the one-particle Hamiltonian level. Figure 2.5 illustrates the band structure of *trans*-polyacetylene and poly(*para*-phenylene vinylene) derived from *ab initio* calculations. Evidently the valence (bonding) and conduction (antibonding) π bands (denoted as π and π^*) are separated from the σ bonding and antibonding bands. The low energy electronic transitions of $\sim 2 - 3$ eV are $\pi \to \pi^*$ excitations, while the $\sigma \to \sigma^*$ excitations lie much higher in energy, being greater than 10 eV. It is this convenient separation of energy scales that means that π-electron models provide an accurate representation of the low-energy physics of conjugated molecules.

2.6 π-electron models

Figure 2.6 is a schematic representation of a conjugated polymer when described by a π-electron model. Each vertex is a site representing a C-H group with one π-orbital. On average, there is one electron per π-orbital.

In π-electron models the σ and core electrons play the static role of screening the Coulomb interactions between the remaining degrees of freedom. In particular, they screen the nuclear-nuclear interactions, the interactions between the π-electrons and the nuclei, and the mutual interactions between the π-electrons. This screening is often modelled by a static dielectric constant,[4] and by the reduction of the effective charge of the nucleus to $+Q$ at large distances. We now define $V_p(\mathbf{r}; \{\mathbf{R}\})$ as the pseudopotential which models the effective interaction between the π-electrons and the nuclei, while $V_{e-e}^{\text{eff}}(\mathbf{r} - \mathbf{r}')$ models the effective electron-electron interaction.

Now, the second quantization representation of the Born-Oppenheimer Hamiltonian, eqn (2.19), is valid for an orthonormal basis. Since the atomic orbitals are not automatically orthonormal, they must first be orthogonalized before they are ready for use. Then we define,

$$\tilde{t}_{ij} = \int \phi_i^*(\mathbf{r}) \left[\frac{\mathbf{p}^2}{2m} + V_p(\mathbf{r}; \{\mathbf{R}\}) \right] \phi_j(\mathbf{r}) d^3\mathbf{r} \tag{2.26}$$

and

$$\tilde{V}_{ijkl} = \int \int \phi_i^*(\mathbf{r}) \phi_k^*(\mathbf{r}') V_{e-e}^{\text{eff}}(\mathbf{r} - \mathbf{r}') \phi_l(\mathbf{r}') \phi_j(\mathbf{r}) d^3\mathbf{r} d^3\mathbf{r}'. \tag{2.27}$$

The four-centre integrals, $\hat{V}_{ijkl}$, are dominated by the diagonal terms:

---

[4]The refractive index of saturated molecules is typically 1.5, implying an effective, static dielectric constant from the σ electrons of ≈ 2.25.

$$U_i \equiv \tilde{V}_{iiii} = \int \int \phi_i^*(\mathbf{r})\phi_i(\mathbf{r})V_{e-e}^{\text{eff}}(\mathbf{r}-\mathbf{r}')\phi_i^*(\mathbf{r}')\phi_i(\mathbf{r}')d^3\mathbf{r}d^3\mathbf{r}', \qquad (2.28)$$

which is the interaction between electrons in the same orbital, and

$$V_{ij} \equiv \tilde{V}_{iijj} = \int \int \phi_i^*(\mathbf{r})\phi_i(\mathbf{r})V_{e-e}^{\text{eff}}(\mathbf{r}-\mathbf{r}')\phi_j^*(\mathbf{r}')\phi_j(\mathbf{r}')d^3\mathbf{r}d^3\mathbf{r}', \qquad (2.29)$$

which is the interaction between electrons in orbitals ϕ_i and ϕ_j. Other terms that are sometimes considered are the exchange interaction,

$$K_{ij} \equiv \tilde{V}_{ijji} = \int \int \phi_i^*(\mathbf{r})\phi_j(\mathbf{r})V_{e-e}^{\text{eff}}(\mathbf{r}-\mathbf{r}')\phi_j^*(\mathbf{r}')\phi_i(\mathbf{r}')d^3\mathbf{r}d^3\mathbf{r}', \qquad (2.30)$$

the density-dependent hopping, $\tilde{V}_{iii,i+1}$, and the bond-charge repulsion, $\tilde{V}_{i,i+1,i+1,i}$. These terms are smaller than the diagonal ones, as they are determined by the wave function overlap, $\phi_i^*(\mathbf{r})\phi_j(\mathbf{r})$. Moreover, since in practice π-electron models are parametrized to fit experiment, these terms are generally neglected, and this will be the approach largely adopted in this book.[5]

Another simplification is to assume that the two-centre integrals, t_{ij}, are only nonzero for electrons in the same orbital or on neighbouring orbitals. Thus,

$$\epsilon_i \equiv \tilde{t}_{ii} \qquad (2.31)$$

is the on-site potential energy, and

$$t_i \equiv -\tilde{t}_{i,i+1} \qquad (2.32)$$

is the nearest-neighbour hybridization (or transfer) integral.[6]

With these approximations, we may write a highly simplified Born-Oppenheimer Hamiltonian for the π-electrons as,

$$H_{\text{BO}}^{\pi} = \sum_i \epsilon_i N_i - \sum_{i\sigma} t_i \left(c_{i+1\sigma}^\dagger c_{i\sigma} + c_{i\sigma}^\dagger c_{i+1\sigma}\right)$$
$$+U\sum_i N_{i\uparrow}N_{i\downarrow} + \frac{1}{2}\sum_{i\neq j}V_{ij}N_iN_j + V_{n-n}, \qquad (2.33)$$

where $N_i = \sum_\sigma N_{i\sigma}$ counts the number of electrons in the orbital $\phi_i(\mathbf{r})$ and V_{n-n} is the screened nuclear-nuclear interaction.

Since this Hamiltonian is widely used throughout this book, we now describe the physical meaning of each term.

---

[5]The neglect of the off-diagonal four-centre integrals is either called the Complete Neglect of Differential Overlap (CNDO), or Zero Differential Overlap (ZDO). Models that retain some other four-centre integrals are termed Intermediate Neglect of Differential Overlap (INDO).

[6]It is convenient to define the hybridization integral as positive-definite.

- $\epsilon_i N_i$ is the potential energy of the electrons in the orbital $\phi_i(\mathbf{r})$.
- $-t_i \left(c_{i+1\sigma}^\dagger c_{i\sigma} + c_{i\sigma}^\dagger c_{i+1\sigma} \right)$ represents the transfer of an electron with spin σ between the spin-orbitals $\chi_i(\mathbf{r}, \sigma)$ and $\chi_{i+1}(\mathbf{r}, \sigma)$ with an energy $-t_i$. As the π-orbitals are on different sites, this represents the transfer of electrons from site to site.
- $U N_{i\uparrow} N_{i\downarrow}$ is the Coulomb interaction between two electrons in the same spatial orbital (that is, on the same site).
- $V_{ij} N_i N_j$ is the Coulomb interaction between the electrons in orbital $\phi_i(\mathbf{r})$ and the electrons in orbital $\phi_j(\mathbf{r})$ (that is, on different sites).

Although eqn (2.33) represents a highly simplified model of conjugated molecules, it still remains a considerable challenge to solve, understand and predict its physical behaviour. We discuss various additional approximations to H_{BO}^π in Section 2.8. However, in the next section we discuss going beyond the Born-Oppenheimer approximation to include explicit electron-phonon coupling.

2.7 Electron-phonon coupling

To derive a simple model of electron-phonon coupling, let us expand the π-electron-nuclear interaction, $V_p(\mathbf{r}; \{\mathbf{R}\})$, around some *reference* set of coordinates, $\{\mathbf{R}^0\}$:

$$V_p(\mathbf{r}; \{\mathbf{R}\}) = V_p(\mathbf{r}; \{\mathbf{R}^0\}) + \sum_l \frac{\partial V_p}{\partial \mathbf{R}_l} \cdot \mathbf{u}_l + \cdots , \tag{2.34}$$

where $\mathbf{u}_l$ is the displacement of the lth ion from its reference position. We define,

$$\epsilon_i^0 = \int \phi_i^*(\mathbf{r}) \left[\frac{\mathbf{p}^2}{2m} + V_p(\mathbf{r}; \{\mathbf{R}^0\}) \right] \phi_i(\mathbf{r}) d^3\mathbf{r}, \tag{2.35}$$

$$t_i^0 = -\int \phi_i^*(\mathbf{r}) \left[\frac{\mathbf{p}^2}{2m} + V_p(\mathbf{r}; \{\mathbf{R}^0\}) \right] \phi_{i+1}(\mathbf{r}) d^3\mathbf{r}, \tag{2.36}$$

and

$$\alpha_{mnl} = \int \phi_m^*(\mathbf{r}) \left[\frac{\partial V_p(\mathbf{r}; \{\mathbf{R}^0\})}{\partial \mathbf{R}_l} \right] \phi_n(\mathbf{r}) d^3\mathbf{r}. \tag{2.37}$$

Then, if we define

$$\alpha \equiv \alpha_{l,l+1,l} = \alpha_{l+1,l,l} = -\alpha_{l,l+1,l+1} = -\alpha_{l+1,l,l+1} \tag{2.38}$$

and

$$\beta \equiv \alpha_{l,l,l+1} = -\alpha_{l,l,l-1} \tag{2.39}$$

we obtain the electron-phonon interaction,

$$H_{e-n}^\pi = \sum_{i\sigma} \left\{ -\alpha \cdot (\mathbf{u}_i - \mathbf{u}_{i+1}) \left(c_{i\sigma}^\dagger c_{i+1\sigma} + c_{i+1\sigma}^\dagger c_{i\sigma} \right) + \beta \cdot (\mathbf{u}_{i+1} - \mathbf{u}_{i-1}) N_{i\sigma} \right\}. \tag{2.40}$$

These terms have a simple, physical interpretation. The first term on the right-hand side is the change in the electronic kinetic energy arising from the changes in the bond lengths from their reference values. Similarly, the second term is the change in the electronic potential energy arising from the changes in bond lengths. Notice that a reduction in the bond lengths results in a decrease of the (negative) kinetic energy. Physically, this is caused by the increase in the magnitude of the negative hybridization integrals as the distances between neighbouring atoms decreases.

The nuclear-nuclear interactions are modelled by the Hamiltonian,

$$H_{n-n} = \sum_{\alpha} \frac{P_{\alpha}^2}{2M_{\alpha}} + V_n(\{\mathbf{u}_n\}), \qquad (2.41)$$

where $V_n(\{\mathbf{u}_n\})$ is the nuclear-nuclear potential energy associated with small displacements from the reference coordinates.

The next step is to *quantize* the nuclear degrees of freedom as phonons, giving a fully quantum mechanical description of the electron and nuclear degrees of freedom. This step will be described in Chapter 7.

2.7.1 The nuclear-nuclear potential, $V_n(\{\mathbf{u}_n\})$

We conclude this section by making some remarks on the nuclear-nuclear potential. It is convenient to separate this into an effective nuclear-nuclear potential arising from the nuclear charges associated with the σ bonds, V_n^{σ}, and the nuclear-nuclear potential from the remaining unscreened nuclear charges associated with the π electrons, V_n^{π}:

$$V_n(\{\mathbf{u}_n\}) = V_n^{\sigma} + V_n^{\pi}. \qquad (2.42)$$

If we suppose that the reference structure is determined by the σ bonding alone, as for example in polyethylene, and that distortions from this structure are small, then we may express V_n^{σ} as a sum of harmonic springs,

$$V_n^{\sigma} = \frac{K}{2} \sum_i (\mathbf{u}_{i+1} - \mathbf{u}_i)^2. \qquad (2.43)$$

K is therefore the spring constant associated with the σ-bonds.

Now suppose that we consider the molecular structure arising from both the σ and π electrons, as for example in *trans*-polyacetylene. As we shall see shortly, the coupling of the π electrons to the lattice leads to both an overall reduction in the chain length, and to a regular distortion of the lattice. Since we want to describe the regular distortion relative to the *average* bond length, r_0, rather than the *reference* bond length, r_e, (determined by the σ electrons), it is convenient to expand V_n^{σ} about r_0. Then,

$$V_n^{\sigma} = \frac{K}{2} \sum_i (\mathbf{u}_{i+1} - \mathbf{u}_i)^2 + K\delta r \sum_i (\mathbf{u}_{i+1} - \mathbf{u}_i) - \frac{NK\delta r}{2}, \qquad (2.44)$$

where δr is the average change in bond lengths caused by the π-electrons:

$$\delta r = r_0 - r_e < 0. \tag{2.45}$$

Using the Hellmann-Feynman theorem, we will show in Sections 4.4 and 7.2 that,

$$\delta r = -\frac{\left(2\alpha\overline{\langle\hat{T}\rangle} - \beta\overline{\langle\hat{D}\rangle}\right)}{K}, \tag{2.46}$$

where $\hat{T}$ is the bond order operator, defined in eqn (4.10), $\hat{D}$ is the bond density-density correlator, defined in eqn (7.11). The overbar represents a spatial average, and α and β are the electron-phonon coupling parameters, defined in eqns (2.38) and (2.39).

Hereafter, we adopt eqn (2.44) as the elastic potential energy resulting from the σ bonds. We include the linear term, but neglect the final constant term.

2.8 Summary of π-electron models

Even though the neglect of the σ electronic dynamics leads to considerable simplifications, the full π-electron-nuclear Hamiltonian is still too complicated to solve exactly. In this section we introduce various approximations to the complete Hamiltonian that make it more tractable. These models are summarized in Table 2.1. Their physical properties are discussed in more detail in the chapters that follow, as indicated in Table 2.1.

We also emphasize that even if accurate effective potentials could be derived from first principles, their utility would be limited because of the errors associated with the neglect of the dynamical influences of the σ and core electrons. In practice, therefore, *semiempirical* parameters are often used, which are derived by fitting the predictions of model Hamiltonians to some known experimental results. This is often a very successful procedure.

2.8.1 The Hückel model

The most drastic approximation is to fix the positions of the nuclei and to neglect the electron-electron interactions. Noninteracting electrons with fixed nuclei geometry are described by the Hückel model (Hückel 1931, 1932), defined as

$$H = \sum_i \epsilon_i N_i - \sum_{i\sigma} t_i \left(c_{i+1\sigma}^\dagger c_{i\sigma} + c_{i\sigma}^\dagger c_{i+1\sigma}\right), \tag{2.47}$$

where we define δ_i such that $t_i = t(1 + \delta_i)$. δ_i is therefore the relative distortion of the ith bond from its average value, where positive and negative values correspond to shortened and lengthened bonds, respectively. Equation (2.47) is the Born-Oppenheimer Hamiltonian, eqn (2.33), with $U = V = 0$, and neglecting the constant V_{n-n} term.

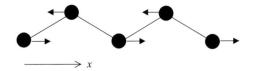

FIG. 2.7. The bond stretching mode of *trans*-CH$_x$ projected onto the x-axis.

2.8.2 *The Su-Schrieffer-Heeger model*

In a π-electron theory the 'ion' represents the CH group, so there are three ionic degrees of freedom per unit cell. These ionic degrees of freedom may be formally represented as collective, normal modes. Su, Schrieffer, and Heeger, in their treatment of *trans*-polyacetylene, introduced a simplification to this problem (Su *et al.* 1970). This simplification was to consider only the normal mode that predominantly couples to the π-electrons. For polyacetylene this is the carbon bond stretching vibration. Thus, projecting the ionic coordinates onto the chain axis, denoted by the x-axis, we have the Su-Schrieffer-Heeger (SSH) Hamiltonian, defined as,

$$H_{\text{SSH}} = H_e + H_{n-n} + H_{e-n}, \tag{2.48}$$

where

$$H_e = -t \sum_{i\sigma} \left(c_{i+1\sigma}^{\dagger} c_{i\sigma} + c_{i\sigma}^{\dagger} c_{i+1,\sigma} \right), \tag{2.49}$$

$$H_{n-n} = \sum_i \left(\frac{P_{i,x}^2}{2M} + \frac{K_x}{2}(u_{i+1,x} - u_{i,x})^2 + K_x \delta r(u_{i+1,x} - u_{i,x}) \right), \tag{2.50}$$

and

$$H_{e-n} = -\sum_{i\sigma} \alpha_x (u_{i,x} - u_{i+1,x}) \left(c_{i+1,\sigma}^{\dagger} c_{i,\sigma} + c_{i,\sigma}^{\dagger} c_{i+1,\sigma} \right). \tag{2.51}$$

We have set $\beta = 0$ (eqn (2.39)) and $\epsilon_i = 0$. Assuming that the bond angles remain fixed at 120^0 during the bond stretching, the projected parameters are defined as, $u_{n,x} = 2u_n/\sqrt{3}$, $\alpha_x = \sqrt{3}\alpha/2$ and $K_x = 3K/4$ (Baeriswyl 1985). The dynamics of the Su-Schrieffer-Heeger model are shown in Fig. 2.7.

The Su-Schrieffer-Heeger model in the limit of static nuclei is known as the *Peierls* model. This is defined and discussed in Section 4.2.

2.8.3 *The Pariser-Parr-Pople model*

Interacting electrons with fixed nuclei satisfy the Pariser-Parr-Pople model (Pariser and Parr 1953a, b; Pople 1953, 1954),[7] defined as

---

[7]The Pariser-Parr-Pople model is known as the extended Hubbard model in solid state physics.

$$H = -\sum_{i\sigma} t_i(c_{i\sigma}^\dagger c_{i+1\sigma} + c_{i+1\sigma}^\dagger c_{i\sigma}) \tag{2.52}$$

$$+ U\sum_i \left(N_{i\uparrow} - \frac{1}{2}\right)\left(N_{i\downarrow} - \frac{1}{2}\right) + \frac{1}{2}\sum_{i\neq j} V_{ij}(N_i - 1)(N_j - 1).$$

The use of the term $(N_i - 1)$ in the Coulomb interactions ensures that this Hamiltonian automatically contains the electron-nuclear and nuclear-nuclear interactions from the nuclear charges associated with the π-electrons. To see this, let us expand the Coulomb interaction:

$$\frac{1}{2}\sum_{i\neq j} V_{ij}(N_i - 1)(N_j - 1) = \frac{1}{2}\sum_{i\neq j} V_{ij}N_iN_j + \sum_{i\neq j} V_{ij}N_j + \frac{1}{2}\sum_{i\neq j} V_{ij} \tag{2.53}$$

The first term on the right-hand side is the electron-electron Coulomb interaction. The second term is the potential energy experienced by the electrons from the nuclei, where

$$\tilde{V}_j = \sum_i V_{ij} \tag{2.54}$$

is the potential energy on site j.[8] Finally, the third term is the nuclear-nuclear potential energy, V_n^π, of eqn (2.42).

The electron-electron interactions are usually treated using the semiempirical Ohno or Mataga-Nishimoto potentials. These expressions are interpolations between a Coulomb potential, $e^2/4\pi\epsilon_0 r_{ij}$, at large separations and U for the interaction between two electrons in the same orbital ($r_{ij} = 0$). For bond-lengths in $\mathring{A}$ and energies in eV the Ohno potential is

$$V_{ij} = \frac{U}{\sqrt{1 + (U\epsilon r_{ij}/14.397)^2}}, \tag{2.55}$$

and the Mataga-Nishimoto potential is

$$V_{ij} = \frac{U}{1 + U\epsilon r_{ij}/14.397}. \tag{2.56}$$

ϵ is the dielectric function, which is usually set to unity. Typically, U is taken to be the value of the ionization potential minus the electron affinity, which is ca. 11 eV in conjugated molecules.

2.9 Symmetries and quantum numbers

Most linear conjugated molecules and polymers possess spatial symmetries, while cyclic polymers possess axial symmetry. Conjugated systems also possess an approximate particle-hole symmetry. These symmetries characterize the electronic

---

[8]Note that the potential energy is also $\tilde{V}_j = \int \phi_j^*(\mathbf{r})V_p(\mathbf{r}; \{\mathbf{R}^0\})\phi_j(\mathbf{r})d^3\mathbf{r}$, where $V_p(\mathbf{r}; \{\mathbf{R}^0\})$ is the pseudopotential defined in Section 2.6.

Table 2.1 *Summary of π-electron models*

Model	Comments	Chapter(s)
Hückel	Noninteracting electrons with a fixed geometry	3
Su-Schrieffer-Heeger (SSH)	Noninteracting electrons with dynamic nuclei	4
Peierls	Static-nuclear limit of the SSH model	4
Pariser-Parr-Pople (P-P-P)	Interacting electrons with a fixed geometry	5, 6
P-P-P-SSH	Interacting electrons with dynamic nuclei	7
Pariser-Parr-Pople-Peierls	Static-nuclear limit of the P-P-P-SSH model	7

states and determine whether or not the states are optically active. We introduce these symmetries here, deferring a full discussion of the physical significance of particle-hole symmetry to Chapter 3.

2.9.1 *Spatial symmetries*

As Figs 1.1 and 1.2 illustrate for *trans*-polyacetylene, poly(*para*-phenylene vinylene), and polydiacetylene, most conjugated polymers possess a two-fold rotation symmetry about an axis of symmetry through their centre and normal to their plane of symmetry. Such polymers are said to possess C_{2h} symmetry (Atkins and Friedman 1997).

C_{2h} symmetry is equivalent to inversion symmetry defined by

$$\mathbf{r} \mapsto -\mathbf{r}. \tag{2.57}$$

A many body state that is even under inversion (with a positive eigenvalue) is denoted A_g, while a many body state that is odd under inversion (with a negative eigenvalue) is denoted B_u.

As illustrated in Fig. 1.2 poly(*poly*-phenylene), possesses planes of symmetry through both the major and minor axes. This is denoted as D_{2h} symmetry. The character table for D_{2h} symmetry is shown in Table 11.3.

2.9.2 *Particle-hole symmetry*

If a Hamiltonian has particle-hole (or charge-conjugation) symmetry then it is invariant under the transformation of a particle into a hole under the action of the particle-hole operator, $\hat{J}$:

$$c_{i\sigma}^{\dagger} \mapsto (-1)^i c_{i\bar{\sigma}} \equiv (-1)^i h_{i\sigma}^{\dagger}, \tag{2.58}$$

where $h_{i\sigma}^{\dagger}$ creates a hole with spin σ, and $\bar{\sigma}$ means the opposite spin to σ.

There are two requirements for an interacting model to posses particle-hole symmetry. The first requirement applies to the kinetic energy, and states that the lattice must be composed of two interpenetrating sublattices, with nearest neighbour one-electron hybridization between the two sub-lattices. As shown in

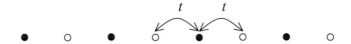

FIG. 2.8. A linear chain with particle-hole, or charge-conjugation, or alternacy symmetry. The lattice is composed of two interpenetrating sublattices (shown as open and filled circles), with nearest neighbour hybridization between the sublattices.

Fig. 2.8, this requirement is satisfied for a one-dimensional chain with nearest neighbour hybridization. As a consequence of particle-hole symmetry the kinetic energy for a uniform cyclic chain satisfies $\epsilon_k = -\epsilon_{k+\pi/a}$, as shown in Fig. 3.1. Similarly, for a linear uniform chain the kinetic energy satisfies $\epsilon_j = -\epsilon_{j-\pi}$, as shown in Fig. 3.2. Thus, in both cases, the energy spectrum is symmetric about $\epsilon = 0$.

The second requirement for a model to posses particle-hole symmetry is that the electron-electron interactions must be balanced - on average - by electron-nuclear interactions. For a chain with translational symmetry every site is equivalent with the same potential energy. For a linear chain, with open boundary conditions, however, the sites are not equivalent. The electrons on sites in the middle of a chain experience a larger potential energy from the nuclei than electrons on sites towards the ends of the chain. This potential energy, $\tilde{V}_j = \sum_i V_{ij}$, is shown in Fig. 2.9. Correspondingly, the electrons on sites in the middle of the chain experience a larger electron-electron repulsion than electrons towards the end of the chain. When this repulsion is equal and opposite to the electron-nuclei attraction, there is particle-hole symmetry, and every site is *essentially* equivalent.

It is easy to demonstrate that the kinetic energy term is invariant under the transformation, eqn (2.58). Also, because under the particle-hole transformation

$$(N_i - 1) \rightarrow -(N_i - 1) \tag{2.59}$$

the Coulomb interactions in the Pariser-Parr-Pople model are also invariant under this transformation. Particle-hole symmetry is an exact symmetry of π-electron models, but since these models are approximate it should be noted that particle-hole symmetry is only an approximate symmetry for conjugated polymers. It is strongly violated in systems with heteroatoms.

Systems which posses particle-hole symmetry satisfy a number of properties. First, the expectation value of the occupancy of each site is unity, or,

$$\langle \hat{N} - 1 \rangle = 0. \tag{2.60}$$

For the π-electron models this means that the average number of π-electrons per site is unity. This result is proved in Appendix B. A second property is that singlet particle-hole excitations that are negative under a particle-hole transformation have an even particle-hole spatial parity, while singlet particle-hole

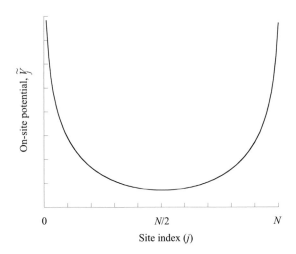

FIG. 2.9. The on-site pseudopotential, $\tilde{V}_j$, of a linear chain of 100 sites, using the Ohno potential, eqn (2.55). On average, this is balanced by the electron-electron repulsion in systems with particle-hole symmetry.

Table 2.2 *States and symmetry character table for linear polymers described by π-electron Hamiltonians with inversion and particle-hole symmetries*

State	Particle-hole eigenvalue	Inversion eigenvalue	Spin
$^1A_g^+$	$+1$	$+1$	0
$^1B_u^-$	-1	-1	0
$^3B_u^+$	$+1$	-1	1
$^3A_g^-$	-1	$+1$	1

excitations that are positive under a particle-hole transformation have an odd particle-hole spatial parity. In contrast, triplet states that are positive under a particle-hole transformation have an even particle-hole parity, and vice versa. This result is proved in Section 3.6.1. It has important consequences for understanding excitons, as we describe in Chapter 6.

2.9.3 *Quantum numbers*

As spin-orbit coupling is weak in conjugated systems the total spin is a conserved quantum number. The low-lying energy eigenstates are singlet ($S = 0$) and triplet ($S = 1$) states.

2.9.4 *State labels*

The many-body states are labelled as $n^p X^\pm$, where

- n is the overall quantum number,
- $p = 1$ for singlets and $p = 3$ for triplets

- $X = A_g$ or B_u for even or odd inversion symmetry, respectively,
- $\pm$ refers to the particle-hole symmetry eigenvalue being ± 1.

Table 2.2 summarizes the symmetries and quantum numbers of the low-lying states of linear molecules.[9] In the noninteracting limit the singlet and triplet $1B_u$ states are degenerate, and lie below the $2A_g$ state. In large part, the aim of this book is to explain how electronic interactions and electron-phonon coupling determines the character and energetic ordering of these states. First, however, we must discuss these states in the noninteracting limit, which is the subject of the following chapter.

---

[9]We show in Chapter 8 that the dipole operator connects states of the same spin with opposite spatial and particle-hole symmetries.

3

NONINTERACTING ELECTRONS

3.1 Introduction

We start our investigation of the electronic states of conjugated polymers by solving the simplest possible model, the noninteracting (or Hückel) model. We apply this to both uniform and dimerized, cyclic and linear polymers. In all cases the polymers have rigid geometries. This procedure naturally introduces the concepts of Bloch and molecular-orbital states, from which the many-body eigenstates are derived. We conclude this chapter by discussing the importance of symmetries on the properties of the particle-hole excitations from the ground state.

The dimerized chain is the simplest model of semiconducting polymers, and is applied in particular to *trans*-polyacetylene. The noninteracting electronic structure of conjugated polymers with more complex unit cells, such as poly(*para*-phenylene), will be discussed in their relevant chapters. We emphasize that the noninteracting model is a simple model. It is not a realistic description of the electronic states of conjugated polymers, as it neglects two key physical phenomena: electron-phonon coupling and electron-electron interactions. Despite these deficiencies it does provide a useful framework for the more complex descriptions to be described in later chapters.

3.2 The noninteracting (Hückel) Hamiltonian

The noninteracting π-model of conjugated polymers introduced in Chapter 2, is,

$$H = - \sum_{n=1,\sigma}^{N} t_n \left(c_{n\sigma}^{\dagger} c_{n+1\sigma} + c_{n+1\sigma}^{\dagger} c_{n\sigma} \right), \tag{3.1}$$

where $t_n = t(1+\delta_n)$ and we have set the on-site energy, $\epsilon_n = 0$. N is the number of sites. For polymers with alternating short and long bonds, $\delta_n = \delta(-1)^n$, where δ_n is positive or negative for short or long bonds, respectively.

3.3 Undimerized chains

We first consider undimerized chains with $\delta = 0$ and $t_n \equiv t$.

3.3.1 Cyclic chains

For periodic boundary conditions eqn (3.1) with $\delta = 0$ is diagonalized by the Bloch transforms,

$$c_{n\sigma}^{\dagger} = \frac{1}{\sqrt{N}} \sum_{k} c_{k\sigma}^{\dagger} \exp(ikna), \qquad (3.2)$$

and

$$c_{n\sigma} = \frac{1}{\sqrt{N}} \sum_{k} c_{k\sigma} \exp(-ikna). \qquad (3.3)$$

The Bloch wavevector, $k = 2\pi j/Na$ and the (angular momentum) quantum number j satisfies, $-N/2 \le j \le N/2$. a is the lattice parameter. The inverse of eqns (3.2) and (3.3) are,

$$c_{k\sigma}^{\dagger} = \frac{1}{\sqrt{N}} \sum_{n} c_{n\sigma}^{\dagger} \exp(-ikna) \qquad (3.4)$$

and

$$c_{k\sigma} = \frac{1}{\sqrt{N}} \sum_{n} c_{n\sigma} \exp(ikna). \qquad (3.5)$$

Substituting eqns (3.2) and (3.3) into eqn (3.1) gives,

$$H = -\frac{t}{N} \sum_{k,k',n} c_{k\sigma}^{\dagger} c_{k'\sigma} \exp(i(k-k')na) \exp(-ik'a) + \text{hermitian conjugate}. \qquad (3.6)$$

Now, using the identity,

$$\frac{1}{N} \sum_{n} \exp(i(k-k')na) = \delta_{kk'}, \qquad (3.7)$$

where $\delta_{kk'}$ is the Kroneker delta-function satisfying,

$$\delta_{kk'} = 1, \text{ if } k = k';$$

and

$$\delta_{kk'} = 0, \text{ if } k \ne k', \qquad (3.8)$$

we have that,

$$H = -2t \sum_{k\sigma} \cos(ka) c_{k\sigma}^{\dagger} c_{k\sigma}. \qquad (3.9)$$

Since $c_{k\sigma}^{\dagger} c_{k\sigma} \equiv N_{k\sigma}$, namely the number operator, we see that H is diagonal in the k-space representation. The single-particle eigenstates are the Bloch states,[10]

$$|k\rangle = c_{k\sigma}^{\dagger}|0\rangle, \qquad (3.10)$$

with eigenvalues,

$$\epsilon_k = -2t \cos(ka). \qquad (3.11)$$

Equation (3.11) is the one-dimensional tight-binding band structure, shown in Fig. 3.1.[11]

---

[10]We drop the spin label when discussing single-particle states.

[11]Notice that there is a two-fold degeneracy in the spectrum, namely $\epsilon_k = \epsilon_{-k}$. This is a consequence of time reversal symmetry, or more correctly, a symmetry in the reversal of

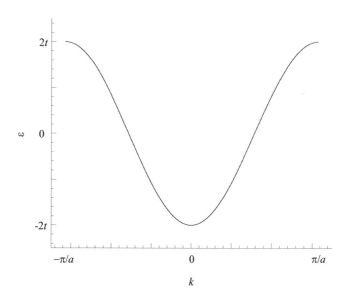

FIG. 3.1. The tight-binding band structure of a cyclic chain, eqn (3.11). As a consequence of particle-hole symmetry, $\epsilon_k = -\epsilon_{k+\pi/a}$, while $\epsilon_k = \epsilon_{-k}$ is a consequence of time reversal invariance.

We may construct the Bloch functions by recalling from Section 2.4 that the creation operator $c_{n\sigma}^{\dagger}$ creates an electron with spin σ in the π-orbital localized on the nth site. Thus, projecting the Bloch state, $|k\rangle$, onto the coordinate representation, $\{|\mathbf{r}\rangle\}$, we have the Bloch function,

$$\psi_k(\mathbf{r}) \equiv \langle \mathbf{r}|k\rangle = \frac{1}{\sqrt{N}} \sum_{n=1}^{N} \phi_n(\mathbf{r}) \exp(-ikna), \qquad (3.12)$$

where we have used eqns (2.14), (2.15), (3.4), and (3.10).

For readers not familiar with the second quantization approach, Appendix A describes a first quantization representation and solution of the eqn (3.1).

3.3.1.1 *The Hückel '4n + 2' rule* The energy spectrum of the cyclic chain explains the Hückel '4n + 2' rule. This rule states that a cyclic chain with N sites is highly stable if $N = 4n+2$, forms a free radical if $N = 4n+1$, and is unstable if $N = 4n$, where n is an integer. The final result follows because filling the energy levels with $4n$ electrons predicts a degenerate ground state. This degeneracy is lifted by a dimerization of the chain; an effect known as the Peierls transition, which we discuss in Chapter 4. The second result simply follows as there are an odd number of electrons. Finally, the first result follows because the ground state

motion. Thus, a rotation (or translation) is equivalent to the reverse rotation (or translation) (Tinkham, 1964).

is nondegenerate, with an energy gap of $O(t/N)$. However, this energy gap $\to 0$ as $N \to \infty$, so even these chains undergo a Peierls transition at a chain-length dependent critical value of the electron-phonon coupling constant. This effect will be discussed more fully in the next chapter.

3.3.2 Linear chains

For linear chains we solve eqn (3.1) using open boundary conditions. The trial solution is

$$c_{n\sigma}^{\dagger} = \sqrt{\frac{2}{N+1}} \sum_{\beta} c_{\beta\sigma}^{\dagger} \sin(\beta na), \qquad (3.13)$$

where the pseudo Bloch wavevector, $\beta = \pi j/(N+1)a$ and j satisfies, $1 \leq j \leq N$. The inverse relation is

$$c_{\beta\sigma}^{\dagger} = \sqrt{\frac{2}{N+1}} \sum_{n} c_{n\sigma}^{\dagger} \sin(\beta na). \qquad (3.14)$$

Substituting eqn (3.13) into eqn (3.1), and using the identity,

$$\frac{2}{N+1} \sum_{n=1}^{N} \sin(\beta na) \sin(\beta' na) = \delta_{\beta\beta'}, \qquad (3.15)$$

leads to the diagonal representation eqn (3.9) with k replaced by β. Thus, the energy of the molecular-orbital state, $|\beta\rangle$, is,

$$\epsilon_{\beta} = -2t \cos(\beta a). \qquad (3.16)$$

This dispersion is shown in Fig. 3.2.

The molecular-orbital functions, $\psi_{\beta}(\mathbf{r})$, are constructed in exact analogy to the Bloch functions of the last section. Thus, we have,

$$\psi_{\beta}(\mathbf{r}) \equiv \langle \mathbf{r}|\beta\rangle = \sqrt{\frac{2}{N+1}} \sum_{n=1}^{N} \phi_{n}(\mathbf{r}) \sin(\beta na). \qquad (3.17)$$

These molecular-orbital functions are particle-in-a-box solutions, and not surprisingly, the molecular-orbital states satisfy the following condition under the operation of the inversion operator $\hat{i}$,

$$\hat{i}|\beta\rangle = i(\beta)|\beta\rangle, \qquad (3.18)$$

where the eigenvalue, $i(\beta)$, is $+1$ for odd quantum number j and -1 for even quantum number j.

3.4 Dimerized chains

The unit cell for a dimerized chain is shown in Fig. 3.3. There are two sites per unit cell, and two different hybridization integrals, $t_d = t(1+\delta)$ and $t_s = t(1-\delta)$, representing the 'double' (short) and 'single' (long) bonds, respectively.

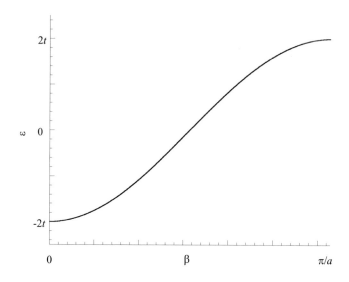

FIG. 3.2. The tight-binding energy spectrum of a linear chain, eqn (3.16). As a consequence of particle-hole symmetry, $\epsilon_\beta = -\epsilon_{\beta - \pi/a}$.

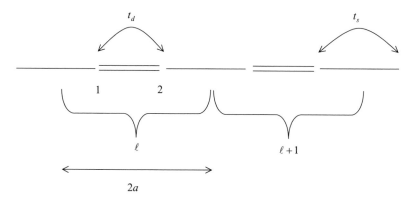

FIG. 3.3. The dimerized chain showing the unit cell and the two sites per unit cell, labelled 1 and 2. t_d and t_s are the 'double' and 'single' bond transfer integrals, respectively. The repeat distance is $2a$.

3.4.1 Cyclic chains

The dimerized chain with periodic boundary conditions is diagonalized by the operators $c_{k\sigma}^{v\dagger}$ and $c_{k\sigma}^{c\dagger}$ that create electrons in Bloch states of the valence and conduction bands, respectively:

$$c_{k\sigma}^{v\dagger} = \frac{1}{\sqrt{2N_u}} \sum_{\ell=1}^{N_u} \left(c_{1\ell\sigma}^{\dagger} \exp(i\chi_k/2) + c_{2\ell\sigma}^{\dagger} \exp(-i\chi_k/2) \right) \exp(-i2\ell ka), \quad (3.19)$$

and

$$c_{k\sigma}^{c\dagger} = \frac{1}{\sqrt{2N_u}} \sum_{\ell=1}^{N_u} \left(c_{1\ell\sigma}^\dagger \exp(i\chi_k/2) - c_{2\ell\sigma}^\dagger \exp(-i\chi_k/2)\right) \exp(-i2\ell ka). \quad (3.20)$$

Here, the sum is over unit cells, $N_u = N/2$, and $k = 2\pi j/Na$, where j satisfies, $-N/4 \leq j \leq N/4$. $c_{1\ell\sigma}^\dagger$ and $c_{2\ell\sigma}^\dagger$ create electrons on sites 1 and 2 of the unit cell ℓ, and

$$\chi_k = \phi_k - ka, \quad (3.21)$$

where

$$\tan(\phi_k) = \delta \tan(ka). \quad (3.22)$$

The corresponding energies are

$$\epsilon_k^v = -2t \left(\cos^2(ka) + \delta^2 \sin^2(ka)\right)^{1/2} \quad (3.23)$$

for the valence band, and

$$\epsilon_k^c = 2t \left(\cos^2(ka) + \delta^2 \sin^2(ka)\right)^{1/2} \quad (3.24)$$

for the conduction band.[12] These results are derived in Appendix C

The band structure is shown in Fig. 3.4. $\epsilon_k = \pm 2t$ at $k = 0$ and $\epsilon_k = \pm 2\delta t$ at $k = \pi/2a$. Thus, the band gap is $4\delta t$, while the full band width is the same as the undimerized chain, namely $4t$. Notice that as δ is increased from 0 to 1 the band structure evolves from that of the undimerized chain (with a folded dispersion), to localized orbitals on the double bonds, with energies of $\pm 2t$.

3.4.1.1 *Wannier States* By Fourier transforming the Bloch operators, $c_{k\sigma}^{\,\,v\,\dagger}$, we obtain Wannier operators, $c_{\ell\sigma}^{\,\,v\,\dagger}$, which create electrons in Wannier states localized on the ℓth repeat unit:

$$c_{\ell\sigma}^{\,\,v\,\dagger} = \frac{1}{N_u} \sum_k c_{k\sigma}^{\,\,v\,\dagger} \exp(i2k\ell a). \quad (3.25)$$

To a rather good approximation,[13] the valence and conduction band Wannier states are equivalent to the bonding and antibonding states, respectively, that is,

$$c_{\ell\sigma}^{\,\,v\,\dagger} \approx \frac{1}{\sqrt{2}} \left(c_{2\ell-1}^\dagger \pm c_{2\ell}^\dagger\right). \quad (3.26)$$

---

[12]For a general t_d and t_s the energy spectrum is $\epsilon_k = \pm \left(t_d^2 + t_s^2 + 2t_d t_s \cos(2ka)\right)^{1/2}$.

[13]The probability amplitude for the Wannier state to overlap a neighbouring dimer is very small. For $\delta = 0.2$ this amplitude is 0.16.

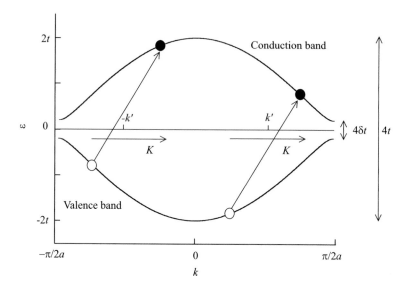

FIG. 3.4. The valence and conduction bands of a dimerized, cyclic chain. The particle-hole excitation at k', and its degenerate counterpart at $-k'$, connected by the
particle-hole transformation, are shown.

3.4.2 Linear chains

The energies of the valence and conduction bands for open, dimerized chains are
again given by eqns (3.23) and (3.24), but with k replaced by β (Lennard-Jones
1937). However, now, unlike the case for undimerized chains, there is no closed
expression for β. Instead, β is determined by the transendental equation,

$$t_d \sin((N+1)\beta a) + t_s \sin(\beta a) = 0. \tag{3.27}$$

Equation (3.27) shows that there is one root in every π/N interval of β for
$\beta = 0 \rightarrow \pi/2a$.[14] The spectrum is plotted in Fig. 3.5.

3.5 The ground state and particle-hole excitations

Once the single-particle eigenstates have been obtained, the many-body states of
the noninteracting Hamiltonian are easily constructed by simply occupying these
single-particle states in accordance with the Pauli exclusion principle. Thus, the
ground state, $|\mathrm{GS}\rangle$, for the dimerized, cyclic chain with one electron, on average,
per π-orbital is found by occupying the valence band:

$$|\mathrm{GS}\rangle = \prod_k c_{k\uparrow}^{v\dagger} c_{k\downarrow}^{v\dagger} |0\rangle. \tag{3.28}$$

---

[14]Equation (3.27) is valid for chains with an even number of sites where the end bonds are
double bonds. If, however, the end bonds are single bonds t_s and t_d are interchanged.

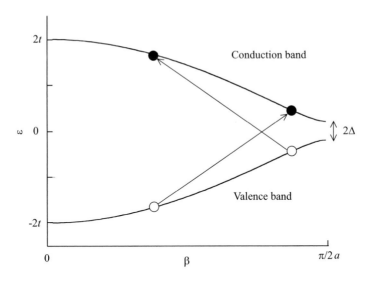

FIG. 3.5. The energy spectrum of the valence and conduction molecular-orbital states for a dimerized, linear chain. A particle-hole excitation and its degenerate counterpart, connected by the particle-hole transformation, are shown. 2Δ is the charge gap, shown as a function of inverse chain length in Fig. 3.6.

The ground state energy, E_0, is thus,

$$E_0 = 2 \sum_k \epsilon_k^v. \tag{3.29}$$

As there are two electrons per Bloch state, the overall total-spin is zero. Such a system is a semiconductor, as there is single-particle gap of $4\delta t$ between the highest occupied valence band state and the lowest unoccupied conduction band state.

An excited state is created by exciting an electron from the valence band to the conduction band, thereby leaving a hole in the valence band,

$$|k_e, k_h\rangle = c_{k_e}^{c\dagger} c_{k_h}^v |\text{GS}\rangle, \tag{3.30}$$

where k_e and $-k_h$ are the wavevectors of the electron and hole, respectively. The total momentum is

$$K = (k_e - k_h), \tag{3.31}$$

and we define the relative momentum as,

$$2k' = (k_e + k_h). \tag{3.32}$$

This excitation is shown in Fig. 3.4. Thus,

$$|k_e, k_h\rangle \equiv |k' + K/2, k' - K/2\rangle. \tag{3.33}$$

The spin label was neglected in eqn (3.30). In fact, a singlet excitation is defined as

$$|^1 k_e, k_h\rangle = \frac{1}{\sqrt{2}} \left(c^{c\dagger}_{k_e \uparrow} c^v_{k_h \uparrow} + c^{c\dagger}_{k_e \downarrow} c^v_{k_h \downarrow} \right) |\text{GS}\rangle, \tag{3.34}$$

while the $S_z = 1, 0$, and -1 triplet excitations are,

$$|^3 k_e, k_h\rangle = c^{c\dagger}_{k_e \uparrow} c^v_{k_h \downarrow} |\text{GS}\rangle, \tag{3.35}$$

$$|^3 k_e, k_h\rangle = \frac{1}{\sqrt{2}} \left(c^{c\dagger}_{k_e \uparrow} c^v_{k_h \uparrow} - c^{c\dagger}_{k_e \downarrow} c^v_{k_h \downarrow} \right) |\text{GS}\rangle, \tag{3.36}$$

and

$$|^3 k_e, k_h\rangle = c^{c\dagger}_{k_e \downarrow} c^v_{k_h \uparrow} |\text{GS}\rangle, \tag{3.37}$$

respectively. The energy of these particle-hole excitations is

$$E(k_e, k_h) = E_0 + \epsilon^c_{k_e} - \epsilon^v_{k_h}, \tag{3.38}$$

with the singlet and triplet states being degenerate. The excitation energy above the ground state is

$$\epsilon(k_e, k_h) = E(k_e, k_h) - E_0 = \epsilon^c_{k_e} - \epsilon^v_{k_h}. \tag{3.39}$$

Exactly the same procedure is employed to construct the many-body-states of the linear chain, except that β replaces k in the above expressions.

3.5.1 The band, charge, and spin gaps

The band gap is the energy between the highest occupied valence band state and the lowest unoccupied conduction band state. This is also the energy of the lowest particle-hole excitation. Now, in a noninteracting model the singlet and triplet excitations are degenerate, so the band gap is equivalent to both the charge and spin gaps. In general we define the charge gap as

$$2\Delta = E_0(N + 1) + E_0(N - 1) - 2E_0(N), \tag{3.40}$$

where $E_0(M)$ is the ground state energy for M electrons. This is obviously equivalent to the band gap, and it is the energy of an *uncorrelated* particle-hole pair.

For short linear chains the charge gap scales linearly with $1/N$, but for long chains it scales as $1/N^2$, approaching $4\delta t$ in the infinite chain limit. This behaviour is shown in Fig. 3.6.[15]

---

[15] It is often erroneously claimed that the single-particle band gap scales as $1/N$ as $N \to \infty$. This is based on the assumption that the single-particle energy levels are particle-in-the-box energy levels, $\epsilon_n = \hbar^2 n^2 / 2ma^2 N^2$, where N is the number of sites. At half-filling, where the number of electrons equals the number of sites, this assumption then predicts that the single particle gap is $\Delta\epsilon = \epsilon_{N/2+1} - \epsilon_{N/2} \sim \hbar^2 / ma^2 N$. As well as for noninteracting systems, as will be shown in Chapter 6, the optical gap in an interacting system also scales as $1/N^2$ as $N \to \infty$.

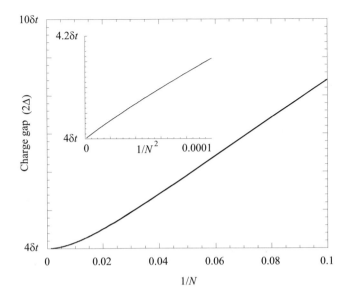

FIG. 3.6. The charge gap, 2Δ, for dimerized, linear chains versus the inverse chain length, and the square of the inverse chain length (inset). As $N \to \infty$ the charge gap scales as $1/N^2$. ($\delta = 0.1$.)

3.6 Symmetries

3.6.1 *Particle-hole symmetry and particle-hole parity*

The particle-hole excitations, defined in Section 3.5, are eigenstates of the noninteracting Hamiltonian, but they are not eigenstates of the particle-hole operator, $\hat{J}$, introduced in Section 2.9.2. To see this, consider the operation of $\hat{J}$ on the singlet excitation, $|^1 k_e, k_h\rangle$:

$$\hat{J}|^1 k_e, k_h\rangle = -\frac{1}{\sqrt{2}} \sum_\sigma c^{c\dagger}_{-k_h \sigma} c^v_{-k_e \sigma} |\mathrm{GS}\rangle$$

$$= -|^1 - k_h, -k_e\rangle \equiv -|^1 - (k' - K/2), -(k' + K/2)\rangle, \quad (3.41)$$

where we have used the relation that under a particle-hole transformation,

$$c^{v\dagger}_{k\sigma} \mapsto -c^v_{-k\bar{\sigma}} \equiv -h^{c\dagger}_{k\sigma}, \quad (3.42)$$

obtained from eqns (2.58), (3.19) and (3.20).[16] Thus, under the particle-hole transformation, $K \mapsto K$ and $k' \mapsto -k'$, as illustrated in Fig. 3.4. We therefore

---

[16]Note that under the particle-hole transformation k and σ are conserved, while the charge is reversed.

see that by forming linear combinations of $|k_e, k_h\rangle$ and $|-k_h, -k_e\rangle$ we create simultaneous eigenstates of H and $\hat{J}$:

$$|K, k'; \mp\rangle = \frac{1}{\sqrt{2}} \left(|k_e, k_h\rangle \pm |-k_h, -k_e\rangle \right), \qquad (3.43)$$

where $|K, k'; -\rangle$ and $|K, k'; +\rangle$ have negative and positive particle-hole symmetry, respectively. In the noninteracting limit these particle-hole states of opposite symmetry are degenerate.

There is an important connection between particle-hole symmetry and the relative parity of the particle-hole pair. Consider a basis state created by the removal of an electron from a valence band Wannier orbital on the repeat unit at $R - r/2$ and the creation of an electron on a conduction band Wannier orbital at $R + r/2$. This is illustrated in Fig. 6.1. This particle-hole pair has a centre-of-mass coordinate, R, and a relative coordinate, r:

$$|R + r/2, R - r/2\rangle = \frac{1}{\sqrt{2}} \sum_\sigma c^{c\dagger}_{R+r/2,\sigma} c^v_{R-r/2,\sigma} |\text{GS}\rangle. \qquad (3.44)$$

Now, using the transformation between Wannier and Bloch states, eqn (3.25), eqn (3.44) can be rewritten as,

$$|R + r/2, R - r/2\rangle =$$

$$\sum_\sigma \frac{1}{\sqrt{N_u}} \sum_{k_e} \frac{1}{\sqrt{N_u}} \sum_{k_h} \exp(i(k_e - k_h)R) \exp(i(k_e + k_h)r/2) c^{c\dagger}_{k_e\sigma} c^v_{k_h\sigma} |\text{GS}\rangle$$

$$= \sum_\sigma \frac{1}{\sqrt{N_u}} \sum_K \frac{1}{\sqrt{N_u}} \sum_{k'} \exp(iKR) \exp(ik'r/2) c^{c\dagger}_{k'+K/2,\sigma} c^v_{k'-K/2,\sigma} |\text{GS}\rangle.$$

$$(3.45)$$

Thus, $|R + r/2, R - r/2\rangle$ is the Fourier transform, with respect to K and k', of $|k' + K/2, k' - K/2\rangle \equiv |k_e, k_h\rangle$. Similarly, the basis state, $|R - r/2, R + r/2\rangle$, obtained by reflecting the electron and hole in eqn (3.44), is the Fourier transform of $|-(k' - K/2), -(k' + K/2)\rangle \equiv |-k_h, -k_e\rangle$. But, $|-(k' - K/2), -(k' + K/2)\rangle$ is connected to $|k' + K/2, k' - K/2\rangle$ by the particle-hole transformation, and thus the linear combination,

$$|R, r; \mp\rangle = \frac{1}{\sqrt{2}} \left(|R + r/2, R - r/2\rangle \pm |R - r/2, R + r/2\rangle \right) \qquad (3.46)$$

is the Fourier transform of the particle-hole adapted state, eqn (3.43). We therefore see that singlet states that are negative under a particle-hole transformation have an even particle-hole parity, while singlet states that are positive under a particle-hole transformation have an odd particle-hole parity. In contrast,

FIG. 3.7. A schematic representation of the phases of the π-orbitals in the HOMO and LUMO, from eqns (3.47) and (3.48).

triplet states that are positive under a particle-hole transformation have an even particle-hole parity, and vice versa.[17] This connection between particle-hole symmetry and the particle-hole parity, which is preserved in interacting models, becomes a useful tool when describing and identifying bound particle-hole, or exciton states, as we will do in Chapter 6.

3.6.2 Linear chains and inversion symmetry

We saw in Section 3.3.2 that the molecular-orbital states of linear chains are eigenstates of the inversion operator, $\hat{\imath}$. The ground state is constructed by occupying each of the valence molecular-orbital states with two electrons. Thus, the overall inversion symmetry of the ground state must be even (or A_g for a many-body state). Now, because the molecular orbital states alternate in symmetry, the highest occupied molecular orbital (HOMO) state will be either even or odd, while the lowest unoccupied molecular orbital (LUMO) state will be either odd or even. In fact, using eqns (3.19) and (3.20), with $\beta = \pi/2a$ replacing k, the HOMO is

$$\psi^{\mathrm{HOMO}}(\mathbf{r}) \equiv \psi^v_{\beta=\pi/2a}(\mathbf{r}) = \frac{1}{\sqrt{2N_u}} \sum_\ell (-1)^\ell \left(\phi_{1\ell}(\mathbf{r}) + \phi_{2\ell}(\mathbf{r}) \right) \qquad (3.47)$$

and the LUMO is

$$\psi^{\mathrm{LUMO}}(\mathbf{r}) \equiv \psi^c_{\beta=\pi/2a}(\mathbf{r}) = \frac{1}{\sqrt{2N_u}} \sum_\ell (-1)^\ell \left(\phi_{1\ell}(\mathbf{r}) - \phi_{2\ell}(\mathbf{r}) \right). \qquad (3.48)$$

The phases of the HOMO and LUMO are shown schematically in Fig. 3.7.

Thus, a particle-hole excitation from the HOMO to the LUMO must have overall odd symmetry. This is the $1B_u$ state. The first A_g excitation (the $2A_g$ state) will be HOMO-1 to LUMO (or, equivalently HOMO to LUMO$+1$). Such an excitation will lie higher in energy than the $1B_u$ state.[18] These transitions are shown in Fig. 3.8.

[17]This is easily proved by using eqn (3.36), and noting the minus sign relative to eqn (3.34).
[18]The energy difference is $O(t/N^2)$, so in the thermodynamic limit the $1B_u$ and $2A_g$ states are degenerate.

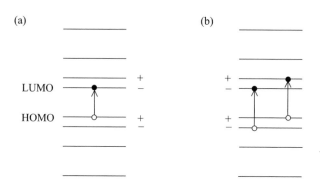

FIG. 3.8. The $1B_u$ (a) and $2A_g$ (b) transitions. The + and − symbols indicate the inversion symmetry of the single-particle molecular-orbital states. The $1B_u$ transition
energy is the charge gap, 2Δ, shown in Fig. 3.6.

As we show in Chapter 8, B_u states are dipole-connected to A_g states, and
thus in the noninteracting model the first excited state decays radiatively to the
ground-state. The charge-gap thus corresponds to the optical gap. The fact that
some conjugated polymers, such as *trans*-polyacetylene, do not electroluminesce
is a consequence of both strong electron-electron and electron-phonon interactions that reverses the energetic ordering of the $1B_u$ and $2A_g$ states. Furthermore,
electron interactions result in an attraction between particle-hole pairs, forming
bound states, or excitons. The excitons lie in the charge gap, and thus, in general, the optical and charge gaps do not coincide. We will return to these points
in later chapters.

4

ELECTRON-LATTICE COUPLING I: NONINTERACTING ELECTRONS

4.1 Introduction

Electron-phonon coupling plays a crucial rôle in one-dimensional systems. For any value of the electron-phonon coupling an infinite, undistorted polymer chain is unstable with respect to a lower symmetry, distorted structure. This is a consequence of the well-known Peierls theorem (Fröhlich 1954; Peierls 1955), which states that a one-dimensional metal is unstable with respect to a lattice distortion that opens a band gap at the Fermi surface. A proof of bond-alternation in conjugated polymers in the noninteracting limit was first presented independently by Ooshika (1957, 1959), and Longuet-Higgins and Salem (1959).

As we describe in this chapter, this mechanism will cause a linear polymer, such as *trans*-polyacetylene, with one π-electron per orbital to have a dimerized ground state composed of alternating short and long bonds. Ooshika (1957) and Longuet-Higgins and Salem (1959) also recognized that a defect in the dimerization, namely a boundary (or domain wall) between one phase of bond-alternation and another (say, long-short-long bonds and short-long-short bonds, as illustrated in Fig. 4.1) is a natural consequence of the broken symmetry ground state. As we will see, there is a fascinating association between these defects and mid-gap electronic states in the semiconducting band gap. This leads to highly mobile unpaired spins, as first predicted by Pople and Walmsley (1962). Many of these early developments are described in (Salem 1966).

In this chapter we describe the consequences of electron-phonon coupling in the absence of electron-electron interactions. The celebrated model for studying this limit is the so-called Su-Schrieffer-Heeger model (Su *et al.* 1979, 1980), defined in Section 2.8.2. In the absence of lattice dynamics this model is known as the Peierls model. We begin by describing the predictions of this model, namely the Peierls mechanism for bond alternation in the ground state and bond defects in the excited states. Finally, we reintroduce lattice dynamics classically and briefly describe amplitude-breathers.

4.2 The Peierls model

It is convenient to define the Peierls model as

$$H_{\text{Peierls}} = H_{\text{kinetic}} + H_{\text{elastic}}, \tag{4.1}$$

where

$$H_{\text{kinetic}} = -2 \sum_n t_n \hat{T}_n \tag{4.2}$$

represents the kinetic energy, and

$$H_{\text{elastic}} = \frac{1}{4\pi t \lambda} \sum_n \Delta_n^2 + \Gamma \sum_n \Delta_n \tag{4.3}$$

is the elastic energy of the σ electrons.

We define $\hat{T}_n$, the *bond order* operator for the nth bond, as

$$\hat{T}_n = \frac{1}{2} \sum_\sigma \left(c_{n+1,\sigma}^\dagger c_{n,\sigma} + c_{n,\sigma}^\dagger c_{n+1,\sigma} \right). \tag{4.4}$$

t_n is the bond hybridization integral,

$$t_n = t + \frac{\Delta_n}{2}, \tag{4.5}$$

where Δ_n is related to the distortion of the nth bond from its average value by

$$\Delta_n = -2\alpha(u_{n+1} - u_n). \tag{4.6}$$

Formally, α is the electron-phonon coupling parameter defined by eqn (2.38), but it is often convenient to regard it as a semiempirical parameter. Notice that a positive value of Δ_n corresponds to a reduction in the bond length, and vice versa. It is this term in t_n that couples the electrons to the lattice, and corresponds to eqn (2.40) with $\beta = 0$. Δ plays the role of an order parameter, whereby a nonzero value indicates a broken symmetry.

H_{elastic} is just V_n^σ - defined by eqn (2.44) (where we have omitted the constant term in eqn (4.3)). Thus, by comparing eqn (4.3) with eqn (2.44) we can define λ, the dimensionless electron-phonon coupling parameter, as

$$\lambda = \frac{2\alpha^2}{\pi K t}, \tag{4.7}$$

and

$$\Gamma = -\frac{K \delta r}{2\alpha}. \tag{4.8}$$

K is the spring constant of the σ-bonds and δr is the average change in bond length relative to the σ-bond reference value due to the π-electrons (see eqn (2.45)).

The expectation value of the bond-order operator is a measure of the strength of that bond, as illustrated by the simple example of ethylene. Modelling this by two π-orbitals with two electrons shared between them it is easily shown that the bonding molecular orbital has a bond-order value of $+1$, while the antibonding molecular orbital has a bond-order value of -1. Thus, a larger bond-order value implies a stronger bond.

We seek a solution of H for arbitrary $\{\Delta_n\}$.[19] In Section 4.4 we discuss the Hellmann-Feynman theorem, which gives us a general solution for any eigenstate. For now, however, we describe the Peierls mechanism, which gives us the dimerized, broken-symmetry ground state.

4.3   The dimerized ground state

Consider a linear, undistorted chain with all $\Delta_n = 0$. Then, the tight binding-band structure is given by eqn (3.11) and shown in Fig. 3.1. Now, if there is on average one π electron per site the Fermi wavevector $k_f = \pi/2a$, where a is the undistorted bond length. Suppose that the the chain dimerizes into long and short bonds so that the unit cell doubles. Then the Brillouin zone will halve in size, and the new Brillouin zone edge will lie at k_f. From standard band theory we know that this will result in a gap opening at k_f, resulting in a reduction of the kinetic energy of the valence electrons. The spectrum of the resulting valence and conduction bands is given by eqns (3.23) and (3.24), and shown in Fig. 3.4. As we now show, this reduction in kinetic energy exceeds the increase in elastic energy that accompanies the distortion. Thus, at half-filling the system spontaneously breaks the discrete translational symmetry and distorts into the lower symmetry dimerized lattice. (Broken symmetries are discussed in general in Section 5.1.1.)

We calculate the equilibrium bond distortion as follows (Longuet-Higgins and Salem 1959). Let us suppose that there is a uniform staggered dimerization,

$$\Delta_n = (-1)^n \Delta. \tag{4.9}$$

Then the total ground state energy is

$$
\begin{aligned}
E_0(\Delta) &= 2 \sum_{k \leq k_f} \epsilon_k^v + \frac{N\Delta^2}{4\pi t \lambda} \\
&= \frac{L}{\pi} \int_{-\pi/2a}^{\pi/2a} \epsilon_k^v dk + \frac{N\Delta^2}{4\pi t \lambda} . \\
&= -\frac{4Nt}{\pi} \int_0^{\pi/2} \left(\cos^2 \theta + \delta^2 \sin^2 \theta \right) d\theta + \frac{Nt\delta^2}{\pi\lambda} ,
\end{aligned}
\tag{4.10}
$$

where $\delta = \Delta/2t$. The first term on the right-hand side of eqn (4.10) is the electronic kinetic energy, while the second term is the elastic energy, and we have used eqn (3.23) for ϵ_k^v.

The energy per site is

$$E_0(\delta)/N = -\frac{4t}{\pi} E(1 - \delta^2) + \frac{t\delta^2}{\pi\lambda} , \tag{4.11}$$

---

[19]Notice that if we divide H by t and define $\tilde{\Delta}_n = \Delta_n/t$ we see that the model is solely parametrized by λ, with t setting the energy scale.

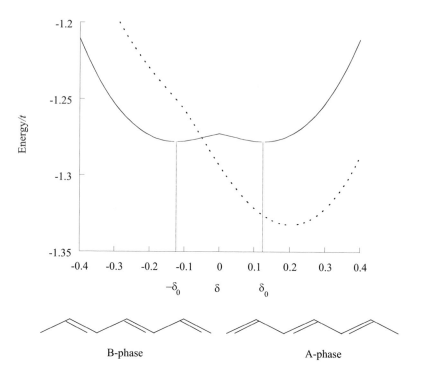

FIG. 4.1. The ground state energy per site as a function of the dimerization parameter, δ. The electron-phonon parameter, $\lambda = 0.2$. (As discussed in Section 4.8, the dashed curve is the ground state energy with an extrinsic bond dimerization, $t_e = 0.1t$.)

where E is the complete elliptical integral of the second kind. For $\delta \ll 1$,

$$E(1 - \delta^2) \approx 1 + \frac{1}{2}\left(\ln\left(4/|\delta|\right) - \frac{1}{2}\right)\delta^2 + \cdots , \qquad (4.12)$$

and hence for small δ,

$$\frac{E_0(\delta)}{N} \approx -\frac{4t}{\pi}\left[1 + \frac{1}{2}\left(\ln\left(4/|\delta|\right) - \frac{1}{2}\right)\delta^2\right] + \frac{t\delta^2}{\pi\lambda}. \qquad (4.13)$$

We see that as a function of δ - for small δ - the kinetic energy decreases more quickly than the increase in elastic energy. Thus, the chain spontaneously dimerizes to a finite value of δ. The energy as a function of δ is shown in Fig. 4.1. Notice that the ground state is doubly degenerate, with δ and $-\delta$ corresponding to the A and B phases, respectively.

Minimizing E_0 with respect to δ we find that there is a saddle point at $\delta = 0$ and stable minima at

$$\delta_0 = \pm 4 \exp\left(-\left[1 + \frac{1}{2\lambda}\right]\right). \qquad (4.14)$$

Thus, the band gap, $2\Delta_0 \equiv 4\delta_0 t$, is

$$2\Delta_0 = 16t \exp\left(-\left[1 + \frac{1}{2\lambda}\right]\right). \qquad (4.15)$$

In the noninteracting limit, the band gap is directly proportional to the dimerization gap. In fact, this prediction is violated in *trans*-polyacteylene, indicating the importance of electron-electron interactions - as we describe in Chapter 7.

The ratio of the band width, $W = 4t$, to the band gap introduces an important concept, namely the coherence length, ξ:

$$\frac{\xi}{a} = \frac{W}{2\Delta_0} = \frac{1}{\delta_0}. \qquad (4.16)$$

4.3.1 The Hückel '4n + 2' rule

The above analysis indicates that the ground state is unstable for an infinitesimally small electron-phonon coupling constant, λ. In fact, this result is only true for linear chains, and for cyclic chains where the number of sites, N, satisfies $N = 4n$, where n is an integer. As discussed in Section 3.3.1.1 the Hückel '4n+2' rule states that cyclic chains where the number of sites satisfies $N = 4n + 2$ are highly stable. For these chains to dimerize the Peierls energy gap, eqn (4.15), must exceed the energy gap of $O(t/N)$ between the highest occupied and lowest unoccupied states of the undimerized chain. This implies that the critical electron-phonon coupling constant, λ_c, satisfies

$$\lambda_c > \frac{1}{2\ln(N)}. \qquad (4.17)$$

Figure 4.2 shows the λ_c versus the inverse chain length for cyclic chains with $N = 4n + 2$.

4.4 Self-consistent equations for $\{\Delta_n\}$

A more general scheme to derive the equilibrium bond distortions, $\{\Delta_n\}$, without resorting to a guess about these distortions, is to require that the force per bond, f_n, is zero.

For the state $|\Psi\rangle$ with eigenvalue E, we define f_n as

$$f_n = -\frac{\partial E}{\partial(u_{n+1} - u_n)} = -\frac{\partial\langle\Psi|H|\Psi\rangle}{\partial(u_{n+1} - u_n)}. \qquad (4.18)$$

Although $|\Psi\rangle$ is a function of $\{\Delta_n\}$, this expression is conveniently evaluated if we use the Hellmann-Feynman theorem, which states that

$$\frac{\partial\langle H(y)\rangle}{\partial y} = \langle\frac{\partial H(y)}{\partial y}\rangle, \qquad (4.19)$$

for any variable y, where $\langle\cdots\rangle$ represents the expectation value with respect to $|\Psi\rangle$ (see Cohen-Tannoudji *et al.* (1977) or Atkins and Friedman (1997) for a proof).

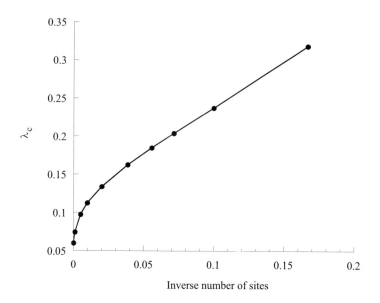

FIG. 4.2. The critical electron-phonon coupling constant, λ_c, for bond alternation to occur versus inverse chain length, N^{-1}, for cyclic chains, where $N = 4n + 2$.

From eqn (4.1) we therefore have,

$$f_n = -2\alpha \left(\frac{\Delta_n}{2\pi t \lambda} + \Gamma - \langle \hat{T}_n \rangle \right). \tag{4.20}$$

Thus, setting $f_n = 0$ gives the following self-consistent equation for Δ_n,

$$\Delta_n = 2\pi t \lambda (\langle \hat{T}_n \rangle - \Gamma). \tag{4.21}$$

When investigating the distortion of the polymer structure by the π-electrons around the *average* bond length, r_0, it is necessary to require a constant chain length, namely,

$$\sum_n (u_{n+1} - u_n) = 0, \tag{4.22}$$

implying that

$$\sum_n \Delta_n = 0. \tag{4.23}$$

Using eqn (4.21), this also implies that

$$\Gamma = \frac{1}{N} \sum_n \langle \hat{T}_n \rangle = \overline{\langle \hat{T}_n \rangle}, \tag{4.24}$$

where the overbar represents the spatial average. Using eqn (4.8) it therefore follows that

$$\delta r = -\frac{2\alpha\overline{\langle \hat{T}_n \rangle}}{K}, \qquad (4.25)$$

confirming eqn (2.46) when $\beta = 0$.

Finally, we now see from eqn (4.21) that the distortion of the nth bond from the average is proportional to the deviation of $\langle \hat{T}_n \rangle$ from its average value, Γ. That is,

$$(u_{n+1} - u_n) = -\frac{\Delta_n}{2\alpha} = -\left(\frac{\pi t \lambda \langle \hat{T}_n \rangle}{\alpha} - \frac{\pi t \lambda \Gamma}{\alpha} \right) \qquad (4.26)$$

$$= -\left(\frac{2\alpha \langle \hat{T}_n \rangle}{K} + \delta r \right).$$

Equation (4.21) can be solved by a numerical iteration scheme. This is particularly useful for excited states where no sensible guess as to $\{\Delta_n\}$ may be possible, and for interacting electron problems where no exact solutions are possible. We discuss interacting electrons and their coupling to phonons in Chapter 7.

We conclude this section by remarking on the character of the broken symmetry ground state. The staggered dimerization of the ground state, represented by eqn (4.9), together with eqn (4.21), implies that there is an alternating deviation of the expectation value of the bond order operator from its average value. This therefore represents a *bond order* wave of strong (short or 'double') and weak (long or 'single') bonds.

4.5 Solitons

4.5.1 *Odd-site chains*

As a result of the degenerate ground state, an immediate and fascinating consequence of bond-alternation are bond-defects, or *solitons*. Solitons separate a dimerized region A from a dimerized region B, and thus they resemble domain walls in ferromagnets.

One way to understand the origin of solitons is to consider an open, linear chain containing an odd number of sites. As in the case of a cyclic chain this linear chain will have a dimerized ground state. Since there are an odd number of sites, there are an even number of bonds. An arrangement of short-long-short-long, etc. bonds (A, say) is degenerate with an arrangement of long-short-long-short, etc. bonds (B, say). However, these arrangements do not form the ground state, because of the end effects that favour short bonds at both ends.[20] There can only be short bonds at both ends if the A phase transforms into the B phase in

---

[20]In principle, the A-phase can tunnel to the B-phase, so a state could lower its energy by being a linear superposition of both phases. However, in the adiabatic approximation this tunnelling is not possible, and even in the nonadiabatic regime the gain in energy is smaller than the loss in energy in linear chains, because of end effects.

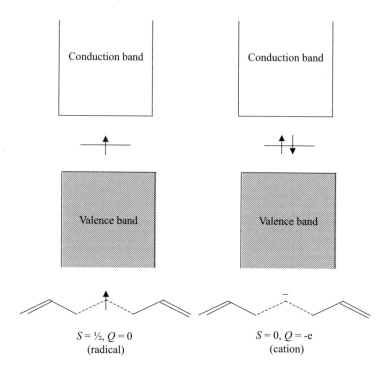

S = ½, Q = 0   S = 0, Q = -e
(radical)   (cation)

FIG. 4.3. The mid-gap state and associated soliton distortions for an odd-site chain.

the middle of the chain via a soliton. The bond defect is shown schematically in Fig. 4.3.

These geometric properties of the chain are also associated with mid-gap states (Pople and Walmsley 1962). To see this, consider the energy spectrum of an even-site chain. There are $N/2$ states in each of the valence and conduction bands. As a result of particle-hole symmetry, every valence band state with energy $\epsilon^v = \epsilon$ maps into a conduction band state with energy $\epsilon_c = -\epsilon$. Thus, the energy spectrum is symmetric about $\epsilon = 0$, as shown in Fig. 3.4. Now, for an odd-site chain there are $(N - 1)/2$ states in each of the valence and conduction bands, and one localized gap state. As a consequence of particle-hole symmetry the localized state lies at $\epsilon = 0$. This mid-gap state is occupied by one electron, and is associated with the soliton, as shown in Fig. 4.3.

By numerically iterating the self-consistent expression, eqn (4.21), the ground state structure of the odd-site chain can be found. Figure 4.4 shows the staggered, normalized bond dimerization, δ_n, defined as

$$\delta_n = (-1)^n \frac{(t_n - \bar{t})}{\bar{t}}, \tag{4.27}$$

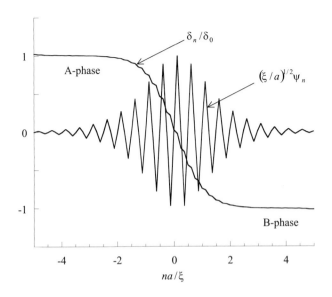

FIG. 4.4. The normalized, staggered bond dimerization, δ_n, of the ground state of an odd-site chain obtained by iterating eqn (4.21). The wavefunction of the mid-gap state, ψ_n, is also shown.

(where $\bar{t}$ is the average value of t_n) for the ground state of an odd-site chain. This curve approximately fits the expression (Su *et al.* 1979),

$$\delta_n = \delta_0 \tanh\left(\frac{(n - n_0)a}{\xi}\right), \tag{4.28}$$

where ξ is the coherence length, defined in eqn (4.16), which determines the width of the soliton centred at n_0.

Also shown in Fig. 4.4 is the single-particle wavefunction, ψ_n, of the mid-gap state, which is localized at the soliton. In the continuum limit, $\xi \gg a$,

$$\psi_n = \left(\frac{a}{\xi}\right)^{1/2} \text{sech}\left(\frac{(n - n_0)a}{\xi}\right) \cos\left(\frac{\pi n}{2}\right). \tag{4.29}$$

An undoped, odd-site chain is charge-neutral with spin 1/2. In this case ψ_n^2 is the *spin* density associated with the soliton. We denote the neutral solitons as S^0. If the chain is doped by one particle, however, the system has charge $\pm e$ and has spin 0, so ψ_n^2 is the *charge* density associated with the soliton. We denote the charged solitons as $S^{\pm}$. The cation is shown schematically in Fig. 4.4. We therefore see that solitons exhibit unusual spin-charge quantum numbers.

4.5.2 *Even-site chains*

A single soliton exists in the ground state of an odd-site chain. For an even-site chain, however, a soliton is paired with an antisoliton so as to restore the

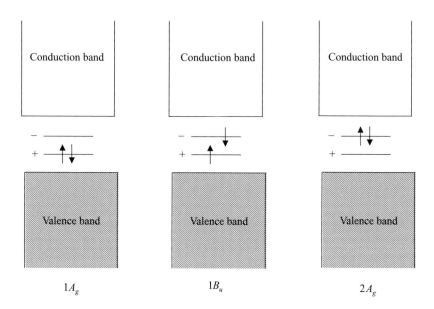

FIG. 4.5. The occupancy of the mid-gap states of the 1^1A_g, $1B_u$, and 2^1A_g states for an even-site chain. The $\pm$ symbols indicate the spatial symmetry of the single-particle states.

bond dimerization. Since solitons change the sign of the staggered dimerization they are known as topological defects. In analogy to the odd-site chain, an even-site chain with a soliton-antisoliton pair has two mid-gap states, symmetrically spaced around $\epsilon = 0$, with energies,

$$\epsilon_0 = \pm\Delta_0 \mathrm{sech}\left(\frac{2n_0 a}{\xi}\right). \tag{4.30}$$

These are single-particle states with a definite spatial symmetry. Thus, the overall spatial symmetry of the many body state is determined by the occupation of these states. Figure 4.5 shows the occupancies for the $1A_g$, $1B_u$, and $2A_g$ states.

Pairs of solitons are the natural excitations from the ground state. This is shown in Fig. 4.6, which shows the bond dimerization of the $1B_u$ state, obtained by iterating eqn (4.21). The bond dimerization fits the functional form (Brazovskii and Kirova 1981; Campbell and Bishop 1981),

$$\delta_n = \delta_0 \left[1 + \tanh\left(\frac{2n_0 a}{\xi}\right) \left\{\tanh\left(\frac{(n - n_0)a}{\xi}\right) - \tanh\left(\frac{(n + n_0)a}{\xi}\right)\right\}\right]. \tag{4.31}$$

We see that there is a soliton at $n = -n_0$, which changes the dimerization from the A to B phase, and an antisoliton at $n = n_0$, which reverses the phase again.

We can use eqn (4.31) to determine the adiabatic energy profiles of the $1A_g$, $1B_u$, and $2A_g$ states as a function of the soliton-antisoliton separation, $R = 2n_0$.

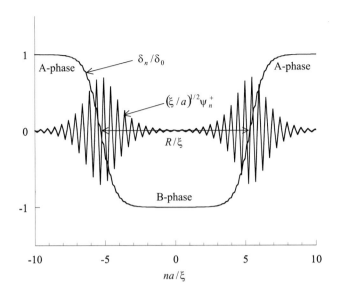

FIG. 4.6. The normalized, staggered bond dimerization, δ_n, of the $1B_u$ state of an even-site chain. The wavefunction of the lower mid-gap state, ψ_n^+, is also shown. (The higher mid-gap state wavefunction, ψ_n^-, has the opposite spatial symmetry.)

This is done by solving eqn (4.1) with the values of $\{\delta_n\}$ from eqn (4.31) for fixed ξ and variable n_0. The energy profiles are shown in Fig. 4.7. We see that the solitons annihilate in the ground state, with the mid-gap states reabsorbed into the valence and conduction bands, leaving a perfectly dimerized chain. In the excited states, however, they repel, and the mid-gap states move to the middle of the gap. For large separations the excitation energies converge to $\sim 4\Delta_0/\pi$, showing that the soliton creation energy is $\sim 2\Delta_0/\pi$.

Solitons are stable in the excited states because of the favourable balance of energies. In the vicinity of the soliton the dimerization is reduced, so the elastic energy is reduced. This reduction in elastic energy more than compensates the increase in kinetic energy associated with localizing the wavefunction near the vicinity of the solitons.

4.6 Soliton-antisoliton pair production

We saw that the soliton in the ground state of an odd-site chain is either neutral with spin-1/2 for the undoped chain (S^0), or charged with spin 0 for the singly doped chain ($S^\pm$). We now discuss the types of solitons present in the excited states of an even-site chain. Suppose that an even-site chain is instantaneously excited from the ground state to the $1B_u$ or $2A_g$ states. This is a *vertical* transition, with the soliton-antisoliton separation initially zero. Within a time $\sim 2\pi/\omega_0$ a soliton-antisoliton pair is created and separates a distance $\sim \xi$. Their

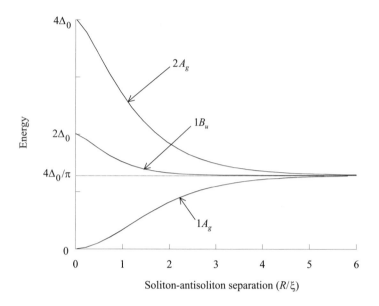

FIG. 4.7. The adiabatic energy profiles of the 1^1A_g, $1B_u$, and 2^1A_g states as a function of soliton-antisoliton separation, R.

trajectory in energy space follows the adiabatic profiles shown in Fig. 4.7. As we now show, as a consequence of the spatial and spin symmetries of the $1B_u$ state, the singlet (1^1B_u) state produces a pair of charged solitons, while the triplet (1^3B_u) state produces a pair of neutral solitons (Ball *et al.* 1983). The $2A_g$ state produces an equal number of charged and neutral pairs.

First, let us consider the singlet, 1^1B_u state. We write this as,

$$|1^1B_u\rangle = \frac{1}{\sqrt{2}} \left(c^\dagger_{+\uparrow}c^\dagger_{-\downarrow} - c^\dagger_{+\downarrow}c^\dagger_{-\uparrow} \right) |V\rangle, \qquad (4.32)$$

where $|V\rangle$ represents the occupied sea of valence states and $c^\dagger_{\pm\sigma}$ creates an electron with spin σ in the mid-gap state $|\psi^\pm\rangle$. As shown in Fig 4.6, the mid-gap wavefunctions, $\psi^\pm_n = \langle n|\psi^\pm\rangle$, resemble the molecular orbitals of a diatomic molecule, as they are linear superpositions of Wannier wavefunctions localized at the centre of each soliton. Denoting the wavefunction localized at the soliton as ϕ_n and the wavefunction localized at the antisoliton as $\overline{\phi}_n$, we have

$$\psi^\pm_n = \frac{1}{\sqrt{2}} \left(\phi_n \pm \overline{\phi}_n \right). \qquad (4.33)$$

For a single soliton, as in the ground state of a odd-site chain, $\phi_n \equiv \psi_n$. This is plotted in Fig. 4.4. The probability distribution functions associated with the localized functions ϕ_n and $\overline{\phi}_n$ are also plotted in Fig. 4.11.

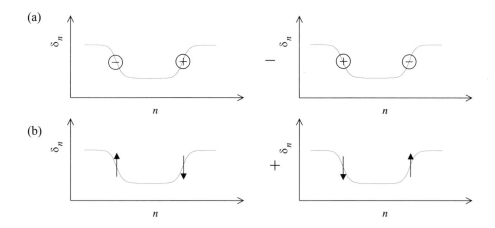

FIG. 4.8. A schematic representation of the (a) 1^1B_u state and (b) the 1^3B_u state, as expressed by eqn (4.35) and eqn (4.36), respectively.

Now, if $c_\sigma^\dagger$ and $\bar{c}_\sigma^\dagger$ creates an electron in the states $|\phi\rangle$ and $|\bar\phi\rangle$, respectively, then

$$c_{\pm\sigma}^\dagger = \frac{1}{\sqrt{2}}\left(c_\sigma^\dagger \pm \bar{c}_\sigma^\dagger\right). \tag{4.34}$$

Inserting eqn (4.34) into eqn (4.32), we have

$$|1^1B_u\rangle = \frac{1}{\sqrt{2}}\left(c_\uparrow^\dagger c_\downarrow^\dagger - \bar{c}_\uparrow^\dagger \bar{c}_\downarrow^\dagger\right)|V\rangle. \tag{4.35}$$

$c_\uparrow^\dagger c_\downarrow^\dagger$ creates a pair of electrons in the soliton, so it is negatively charged and spinless, while the antisoliton contains no electrons, so it is positively charged and also spinless. Similarly, $\bar{c}_\uparrow^\dagger \bar{c}_\downarrow^\dagger$ creates a pair of electrons in the antisoliton, while the soliton contains no electrons. The 1^1B_u state is therefore a linear superposition of spinless positively and negatively charged soliton-antisoliton pairs, as shown schematically in Fig. 4.8(a).

A similar argument applies to the triplet, 1^3B_u state,

$$\begin{aligned}|1^3B_u\rangle &= \frac{1}{\sqrt{2}}\left(c_{+\uparrow}^\dagger c_{-\downarrow}^\dagger + c_{+\downarrow}^\dagger c_{-\uparrow}^\dagger\right)|V\rangle \\ &= \frac{1}{\sqrt{2}}\left(c_\uparrow^\dagger \bar{c}_\downarrow^\dagger + c_\downarrow^\dagger \bar{c}_\uparrow^\dagger\right)|V\rangle,\end{aligned} \tag{4.36}$$

showing that it is a linear superposition of neutral spin-1/2 soliton-antisoliton pairs, as shown schematically in Fig. 4.8(b).

Finally, it is easily shown that the $2^1 A_g$ state, defined as,

$$|2^1 A_g\rangle = c^\dagger_{-1\uparrow} c^\dagger_{-1\downarrow} |V\rangle \tag{4.37}$$

$$= \frac{1}{\sqrt{2}} \left[\frac{1}{\sqrt{2}} \left(c^\dagger_\uparrow \bar{c}^\dagger_\uparrow + c^\dagger_\downarrow \bar{c}^\dagger_\downarrow \right) - \frac{1}{\sqrt{2}} \left(c^\dagger_\uparrow c^\dagger_\downarrow - \bar{c}^\dagger_\uparrow \bar{c}^\dagger_\downarrow \right) \right] |V\rangle$$

has an equal number of charged and neutral soliton-antisoliton pairs.

These results have important consequences for pair production resulting via an optical excitation from the ground state. Since the dipole operator only connects the ground state to $^1 B_u$ states, only charged soliton pairs are produced by this processes. Another interesting consequence, as we discuss in more detail in Chapter 7, is that electron-electron interactions bind the oppositely charged soliton-antisoliton pairs together, creating a strongly bound $1^1 B_u$ exciton. These bound soliton-antisoliton pairs show analogies to polarons, as discussed in the next section, and to confinement of soliton-antisoliton pairs as a result of extrinsic dimerization, as discussed in the following section. In contrast, the neutral solitons of the $1^3 B_u$ and $2^1 A_g$ states couple strongly to the bond order wave, resulting in a significant lattice distortion and energy relaxation for these states.

4.7   Polarons

So far we have mainly focussed on the neutral excitations of the chain. However, nonlinear defects - known as polarons - also exist in the ground state of a doped chain. Polarons are a distortion of the lattice around the doped particle. In the continuum limit the dimerization parameter satisfies the two-soliton expression,

$$\delta_n = \delta_0 \left[1 + \frac{1}{\sqrt{2}} \left\{ \tanh\left(\frac{(n-n_0)a}{\sqrt{2}\xi} \right) - \tanh\left(\frac{(n+n_0)a}{\sqrt{2}\xi} \right) \right\} \right], \tag{4.38}$$

where

$$n_0 = \frac{\xi}{\sqrt{2}a} \ln(1 + \sqrt{2}) \approx 0.623 \frac{\xi}{a}. \tag{4.39}$$

The small separation of the soliton-antisoliton pair means that there is no change of dimerization, merely a reduction in the dimerization in the locality of the doped particle. This is behaviour is shown in Fig. 4.9.

4.8   Nondegenerate systems

Trans-polyacetylene has the unusual property of exhibiting no extrinsic dimerization, and thus has no extrinsic band gap. The dimerization arises entirely from π-electrons coupling to the lattice. Consequently, the A and B phases are degenerate. Most polymers, however, have an extrinsic semiconducting band gap as a result of their stereochemistry independent of the of π-electrons. Examples of polymers that are extrinsically semiconducting include, *cis*-polyacetylene (because of the structure caused by the σ orbitals), polydiacteylene (because of the

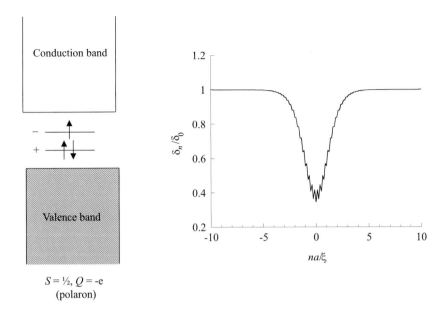

FIG. 4.9. The occupancy of the mid-gap states (left) by a doped particle, and the associated polaronic distortion of the chain (right).

tetramerization caused by the p_y orbitals), and phenyl-based polymers (because of the phenyl rings).

The detailed effects of the stereochemistry vary from polymer to polymer - the details of particular polymers will be described in their relevant chapters. However, to understand the qualitative consequences of extrinsic dimerization we can use the linear chain to model its effects. In this model the π-electrons are coupled to the extrinsic dimerization via the bond integral (Bishop *et al.* 1981; Brazovoskii and Kirova 1981),

$$t_n = t + (-1)^n t_e + \frac{\Delta_n^i}{2} \equiv t + \frac{\Delta_n}{2}, \qquad (4.40)$$

where t_e is the extrinsic bond dimerization and $\Delta_n^i/2$ is the intrinsic dimerization. The kinetic energy of the π-electrons is a function of both the extrinsic and intrinsic dimerizations, while the elastic energy is determined only by the intrinsic dimerization. It is this distinction between the dependence of the kinetic and potential energies on the extrinsic dimerization that causes the nondegenerate ground state, as we discuss shortly.

Assuming a uniform staggered distortion in the ground state, $\Delta_n^i = (-1)^n \Delta^i$, and minimizing the ground state energy, we obtain the self-consistent equation,

$$2\Delta_0 = 4t_e + 2\Delta_0^i = 8t \exp(\gamma) \exp\left(-\left[1 + \frac{1}{2\lambda}\right]\right), \qquad (4.41)$$

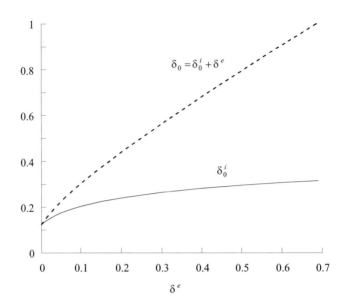

FIG. 4.10. The intrinsic bond dimerization, δ_0^i, and the total bond dimerization, $\delta_0 = \delta_0^i + \delta^e$ as a function of the extrinsic bond dimerization, $\delta^e = t_e/t$. $\lambda = 0.2$.

where

$$\gamma = \frac{t_e}{\lambda \Delta_0}, \tag{4.42}$$

is known as the confinement parameter.

The extrinsic dimerization has two effects. First, it causes an increased intrinsic dimerization, as shown in Fig 4.10. Second, it lifts the degeneracy of the A and B phases, as shown in the plot of the ground state energy in Fig. 4.1. This causes a linear confinement of the soliton-antisoliton pair, because the energy to create a B phase relative to the A phase increases linearly with the soliton-antisoliton separation. This new property of soliton-antisoliton confinement is illustrated by the localized Wannier orbitals associated with the soliton, ϕ_n, and antisoliton, $\bar{\phi}_n$. These are obtained from the molecular orbitals associated with the mid-gap electronic states, $\psi_n^\pm$, (described in Section 4.5) by inverting eqn (4.33). Thus,

$$\phi_n = \frac{1}{\sqrt{2}}(\psi_n^+ + \psi_n^-) \tag{4.43}$$

and

$$\bar{\phi}_n = \frac{1}{\sqrt{2}}(\psi_n^+ - \psi_n^-). \tag{4.44}$$

Figure 4.11 shows the probability density of the Wannier orbitals associated with the mid-gap states. Although the relative separation of Wannier orbitals is small with an extrinsic dimerization of $\delta_e = 0.1$, the fact that there are two

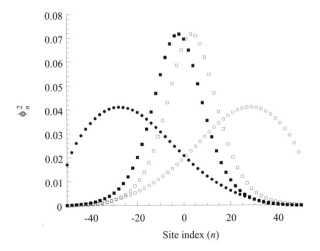

FIG. 4.11. Probability distribution functions of the soliton defects in the noninteracting limit on a 102-site chain for the $1B_u$ state. Left defect, or soliton (filled symbols), right defect, or antisoliton (open symbols); extrinsic dimerization, $\delta_e = 0$ (circles), $\delta_e = 0.1$ (squares) and $\lambda = 0.1$.

distinct Wannier orbitals implies that the argument employed in Section 4.6 - concerning the different characters of the $1^1 B_u^-$ and $1^3 B_u^+$ states after electron-lattice relaxation - is a general one. Thus, the $1^1 B_u^-$ state is comprised of spinless electron-hole pairs, while the $1^3 B_u^+$ state is comprised of two spin-1/2 objects. These become confined in the presence of extrinsic dimerization.

Figure 4.12 shows the soliton-antisoliton pair for various extrinsic dimerizations. We see that even for relatively small extrinsic dimerizations the confinement energy is large enough to prevent a phase reversal between the soliton and antisoliton.

4.9 The continuum limit of the Su-Schrieffer-Heeger model

In the limit that the coherence length is much larger than the lattice spacing, and provided that we are only interested in the low energy physics near to the Fermi surface, a continuum version of the Su-Schrieffer-Heeger model can be derived. This model, derived by Takayama, Lin-Liu, and Maki, is known as the TML model (Takayama *et al.* 1980). It provides useful analytical results that agree with the Su-Schrieffer-Heeger model in the continuum limit.

In the continuum limit, $na \to x$ and $(-1)^n \Delta_n \to \Delta(x)$. Then the TML model is defined as

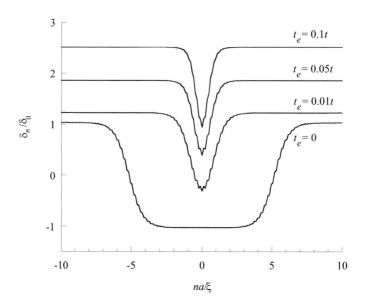

FIG. 4.12. The normalized, staggered bond dimerization, δ_n, of the 1^1B_u state for various extrinsic dimerizations, t_e. (The plotted results are scaled by the $t_e = 0$ values of δ_0 and ξ.) $\lambda = 0.2$.

$$H_{\text{TML}} = \sum_\sigma \int dx \Psi_\sigma^\dagger(x) \left[-i\hbar v_F \frac{\partial}{\partial x} + \Delta(x)\sigma_x \right] \Psi_\sigma(x)$$

$$+ \frac{1}{2\pi\hbar v_F \lambda} \int dx \left[\frac{\dot{\Delta}^2(x)}{\omega_0^2} + \Delta^2(x) \right], \tag{4.45}$$

where $\Psi_\sigma(x)$ are the electron field operators, σ_x and σ_z are the Pauli spin matrices and $v_F = 2ta/\hbar$ is the Fermi velocity. The first term on the right-hand side of eqn (4.45) is the electron kinetic energy in the absence of a bond-order wave. The second term represents the coupling of the electrons to the bond-order density wave, which has the effect of mixing the two components of Ψ_σ on the opposite Fermi points. The final two terms represent the kinetic and elastic energies of the bond-order field.

The uniform, static solution for the bond-order gives,

$$2\Delta_0 = 8t \exp\left(-\frac{1}{2\lambda}\right). \tag{4.46}$$

The soliton defects and wavefunction are given by eqns (4.28) and (4.29), respectively, while the soliton creation energy is $2\Delta_0/\pi$. For further details of the

TML model, we refer the reader to Takayama *et al.* (1980), Baeriswyl (1985), and Heeger *et al.* (1988).

4.10 Dynamics of the Su-Schrieffer-Heeger model

So far in this chapter we have described the static geometrical distortions associated with the electronic states, without paying much regard as to how these distortions arise dynamically. In this section we briefly describe the predicted dynamics of the Su-Schrieffer-Heeger model (introduced in Section 2.8.2).

After photoexcitation or electron-hole injection excess energy in the polymer is liberated during the nonradiative relaxation to lower energy excited states (for example, to the 1^1B_u or 2^1A_g states). Ultimately this energy is lost as heat via the coupling of the intramolecular vibrations to the environment. However, initially it is converted to intramolecular lattice dynamics. A particularly interesting type of lattice dynamics are *breathers* (Su and Schrieffer 1980). A breather, or amplitude-breather, describes localized oscillations in the bond-order amplitude. It may be regarded as a nonlinear excitation of bound phonons resulting from the electron-lattice coupling (Phillpot *et al.* 1989).

Semiclassical solutions of the Su-Schrieffer-Heeger model - whereby the ions are treated classically and are subject to forces determined by the gradients of the adiabatic potential - have been performed for both a degenerate system (namely, *trans*-polyacetylene) and nondegenerate systems. In *trans*-polyacetylene the photoexcited electron-hole pair rapidly diassociates into a widely separated soliton-antisoliton pair, with the excess energy converted into a breather in the centre of the chain (Bishop *et al.* 1984). In nondegenerate systems, however, the soliton-antisoliton pair are confined, and a composite excitation involving the electron, hole and breather develops (Phillpot *et al.* 1989).[21]

4.11 Self-trapping

As Figs 4.4 and 4.6 show, the single-particle wavefunctions of the mid-gap states are localized at the centre of the solitons. In the absence of a driving field the adiabatic approximation predicts that the solitons are static. In other words, the electronic states are trapped at the soliton positions. This is called self-trapping. However, in a translationally invariant system, such self-trapping is an artefact of the approximation. This is because the energy eigenstates should also be eigenstates of the translation operator, which self-trapped states evidently are not. Eigenstates that satisfy this requirement are Bloch states constructed from the basis of the localized (or Wannier) states. In order for the Hamiltonian to connect localized states the lattice dynamics must be restored and treated quantum mechanically. This leads to bands of soliton states. An electronic state

---

[21] The Su-Schrieffer-Heeger model alone is too simplistic to realistically model excited states in conjugated polymers, as electron-electron interactions lead to significantly different predictions. The study of breathers within an interacting electron model has been performed by Takimato and Sasai (1989) and Tretiak *et al.* (2003).

may be regarded as *practically* self-trapped, however, if the inverse bandwidth multiplied by $\hbar$ is longer than experimental observation times. This is equivalent to the statement that if the effective mass becomes so large that the dynamics are slower than observational timescales, then the particle is self-trapped.

A quantum mechanical treatment of phonons and an application to *trans*-polyacetylene will be described in Chapter 10.

4.12 Concluding remarks

This chapter has described the effects of electron-phonon coupling for noninteracting electrons in the adiabatic limit. Bond-alternation and soliton defects in the excited states have been introduced. As stressed in Chapter 1, however, electron-electron interactions also play an important role in determining the electronic properties of conjugated polymers. In some cases the introduction of electron-electron interactions qualitatively changes the predictions of the noninteracting limit; in other cases there are quantitative changes. As an example of a qualitative change, the bond-alternation amplitude is significantly enhanced by electronic interactions. Quantitative changes include the 1^1B_u state changing from being composed of an unbound soliton-antisoliton pair to being an exciton-polaron, and to the 2^1A_g state being composed of a pair of bound soliton-antisoliton pairs. It is also possible for there to be a reversal in the energetic ordering of the 1^1B_u and 2^1A_g states. The effects of electron-electron interactions will be described in the subsequent chapters.

5

INTERACTING ELECTRONS

5.1 Introduction

In this chapter we begin to describe the effects of electron-electron interactions in conjugated polymers. We first discuss broken symmetry ground states before focussing on the character and excitation energies of some of the important low-lying states. A particularly important consequence of electron-electron interactions for neutral one-dimensional systems is the formation of bound particle-hole excitations, or excitons. This subject will be briefly discussed in this chapter, but fuller descriptions of excitons will be given in the next chapter. Electron-phonon interactions will be neglected in this chapter, so throughout we consider π-electron models with fixed geometries. The combined effects of electron-electron and electron-phonon interactions will be described in Chapter 7.

5.1.1 *Broken symmetries*

An electronic state has a broken symmetry if its symmetry is lower than the Hamiltonian that describes it. Broken symmetry ground states occur widely in Nature, for example, in superconductivity, magnetism, and in particle physics (Anderson 1984). Generally, electron systems in one-dimension can exhibit three types of broken symmetries: spin-density waves, charge-density waves, and bond order waves. These waves exhibit particular types of long range correlation, respectively in the spin-density, charge-density, or bond order.

The bond alternation of linear polymers described in Chapter 4 is a bond order wave. The broken symmetry of this ground state is illustrated by its energy shown in Fig. 4.1. The energy is symmetric with respect to the bond alternation parameter, δ. (This reflects the symmetry of the underlying Hamiltonian.) However, there are two equivalent minima at $\pm\delta_0$. If the system can access both configurations at each minima equally, perhaps by quantum tunnelling or thermal activation, then the system has the same symmetry as the Hamiltonian. Conversely, if the system falls into one of these minima and cannot subsequently access the other minimum then the symmetry of the ground state is lower than that of the Hamiltonian. This latter scenario is generally applicable to conjugated polymers.

The periodicity of the correlation is characterized by the structure factor, defined as,

$$S_{\mathrm{SDW}}(q) = \frac{1}{N} \sum_{\ell r} \exp(iqr) \langle (N_{\uparrow\ell} - N_{\downarrow\ell})(N_{\uparrow\ell+r} - N_{\downarrow\ell+r}) \rangle, \qquad (5.1)$$

for the spin-density wave,

$$S_{\text{CDW}}(q) = \frac{1}{N} \sum_{\ell r} \exp(iqr) \langle (N_{\uparrow\ell} + N_{\downarrow\ell})(N_{\uparrow\ell+r} + N_{\downarrow\ell+r}) \rangle, \qquad (5.2)$$

for the charge-density wave, and

$$S_{\text{BO}}(q) = \frac{1}{N} \sum_{\ell r} \exp(iqr) \langle \hat{T}_\ell \hat{T}_{\ell+r} \rangle, \qquad (5.3)$$

for the bond order wave wave. $\hat{T}_\ell$ is the bond order operator for the ℓth bond, defined in in eqn (4.10). The wavevector of the periodicity, q, is related to the Fermi wavevector k_f, via $q = 2k_f$. Thus, a half-filled system exhibits instabilities at $q = \pi/a$ implying a repeat unit of length $2a$.

In an undimerized chain a spin-density wave exhibits gapless spin excitations and gapped charged excitations. However, in a dimerized chain all three types of order exhibit gapped spin and charge excitations. In fact, for a dimerized chain the spin-density and bond order waves coexist. Mazumdar and Campbell have shown (Mazumdar and Campbell 1985) that the Pariser-Pople-Parr model will exhibit a broken-symmetry ground state provided that,

$$V_{i,j+1} - 2V_{i,j} + V_{i,j-1} \geq 0, \qquad (5.4)$$

where $V_{i,j}$ is the Coulomb interaction. They further showed that if,

$$\frac{U}{2} + \sum_j (V_{i,2j} - V_{i,2j-1}) > 0 \qquad (5.5)$$

the bond order (or spin-density) wave is favoured over the charge-density wave, and conversely otherwise. Since the Ohno and Mataga-Nishimoto potentials (and $1/r$ potentials in general) satisfy both conditions we expect that conjugated polymers will generally exhibit bond order (or spin-density) broken symmetry ground states.

5.1.2   Undimerized chains

We now discuss the role of electronic interactions on the electronic spectra of undimerized chains. Electronic interactions via the Coulomb potential have a profound effect on the behaviour of electrons in one dimension. In particular, the usual Fermi liquid behaviour, whereby the interacting electrons are renormalized into weakly interacting quasi-particles that in the metallic state behave as a noninteracting electron gas with a renormalized effective mass, no longer applies. Instead, the electrons are described by a Luttinger liquid, which predicts spin-charge separation, and has quite different transport and thermodynamic properties to a Fermi liquid (Tsvelik 1995; Giamarchi 2003).

Luttinger liquid behaviour applies to metallic systems. However, as already discussed, for a half-filled band the metallic state is unstable with respect to a broken symmetry spin-density wave ground state. There is a gap to charge excitations, and hence the system is an insulator. The spin gap, however, is zero. For the Hubbard model (namely, the Pariser-Parr-Pople model, eqn (2.52), in the limit of only on-site Coulomb interactions) exact results can be obtained for weak and strong interactions (Misurkin and Ovchinnikov 1971; Coll 1974). (See Giamarchi (2003) or Essler *et al.* (2005) for full details on the Hubbard model.)

For weak coupling, $U \ll t$, the charge-excitation (or correlation) gap, 2Δ, is

$$2\Delta = \frac{\sqrt{Ut}}{8\pi} \exp\left(-\frac{2\pi t}{U}\right), \tag{5.6}$$

while for strong coupling, $U \gg t$,

$$2\Delta = U - 4t + \frac{8t^2}{U} \ln(2). \tag{5.7}$$

Such insulating systems are known as Mott-Hubbard insulators. The charge gap separates many-particle states from which electrons can be removed (known as the lower Hubbard band) from many-particle states to which electrons can be added (known as the upper Hubbard band). These bands are quite unlike the single-particle valence and conduction bands described in Chapter 3. For example, as the number of electrons changes the width of the bands and the band gap changes. In principle, the Hubbard bands can be measured experimentally by determining the single-particle spectral weight, $S(\omega)$, defined by

$$S(\omega) = \frac{1}{\pi} \sum_{k\sigma} \text{Im}[G_{k\sigma}^R(\omega)]. \tag{5.8}$$

Here, $G_{k\sigma}^R(\omega)$ is the retarded single-particle Green function,

$$G_{k\sigma}^R(\omega) = \text{Fourier transform}\{\langle \Psi_0 | c_{k\sigma}^\dagger(t) c_{k\sigma}(0) | \Psi_0 \rangle\}, \tag{5.9}$$

where $|\Psi_0\rangle$ is the ground state. In the noninteracting limit, therefore, $S(\omega)$ is the just the single-particle density of states, $\rho(\omega)$. Figure 5.1 shows a schematic diagram of the Hubbard bands in the strong coupling limit.

Since the charge gap is the gap between the highest electron removal state and the lowest electron addition state, we can also define it as

$$2\Delta = E_0(N + 1) + E_0(N - 1) - 2E_0(N), \tag{5.10}$$

where $E_0(M)$ is the ground state energy for M electrons.

Our discussion so far has concentrated on charge excitations. These excitations involve charge transfer from one site to another. We have also restricted our discussion to the Hubbard model. For more realistic models with long range interactions, such as the Pariser-Parr-Pople model (eqn (2.52)), bound electron-hole pairs, or exciton states exist in the single-particle spectral gap. These states will lie an energy equal to their binding energy below the bottom of the upper

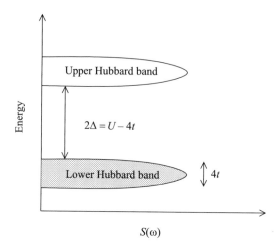

FIG. 5.1. A schematic diagram of the single-particle spectral function, $S(\omega)$, showing the Hubbard bands in the strong-coupling limit, $U \gg 4t$, at half-filling.

Hubbard band. We describe excitons in detail in the next chapter, although a brief description of the energy of bound states in the weak-coupling, undimerized limit is given in Section 5.2.2. For polymers with inversion symmetry the lowest charge-transfer (or ionic) excitation is the $1^1B_u^-$ state. Another kind of excitation from the ground state are spin-density wave (or covalent) excitations. The lowest of these in energy is the spin-one magnon, or the $1^3B_u^+$ triplet state. Pairs of magnons can combine to form singlet states, the lowest in energy being the $2^1A_g^+$ state. In an undimerized chain the spin excitations are gapless, so the $1^3B_u^+$ and $2^1A_g^+$ states always lie below the $1^1B_u^-$ state. For dimerized chains, however, there is a crossover from band transitions to Mott-Hubbard transitions as a function of interaction strength, as we now describe.

5.1.3 *Dimerized chains*

Recall from Chapter 3 that the noninteracting band gap in a dimerized chain is $4\delta t$ (where δ is the dimerization parameter), with the $1^1B_u^-$, $1^3B_u^+$, and $2^1A_g^+$ states being degenerate. For weak electronic interactions these states become bound Mott-Wannier excitons,[22] and their excitation energies increase as a function of the strength of the interactions. However, for stronger interactions the $1^3B_u^+$ and $2^1A_g^+$ states evolve into spin-density-wave states, and their energies begin to decrease (Schulten and Karplus 1972; Tavan and Schulten 1987). The $1^1B_u^-$ state, on the other hand, evolves into a Mott-Hubbard exciton, and its energy eventually increases linearly with U. Figure 5.2 shows the energies of the $1^1B_u^-$, $1^3B_u^+$, and $2^1A_g^+$ states, and the charge gap, 2Δ, as a function of U for

---

[22]Mott-Wannier and Mott-Hubbard excitons are described in the next chapter.

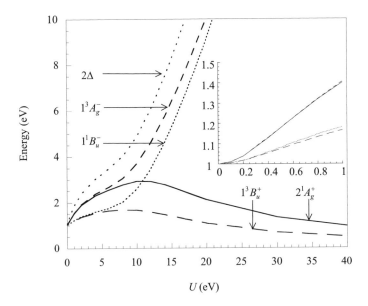

FIG. 5.2. The calculated Pariser-Parr-Pople model excitation energies of the $1^1B_u^-$, $1^3B_u^+$, $1^3A_g^-$, and $2^1A_g^+$ states, and the charge gap, 2Δ, as a function of U on a dimerized chain. This figure illustrates the crossover from band-insulator transitions at small U to Mott-Hubbard transitions at large U. The intermediate parameter regime, when $U \sim 4t$, is applicable to conjugated polymers. $\delta = 0.1$ and $t = 2.5$ eV. The inset shows the excitation energies for small U.

$\delta = 0.1$.

The crossover from band-insulator to Mott-Hubbard insulator occurs in the intermediate-coupling regime, around $U = 4t$. This crossover has been studied by Soos *et al.* (1993), Mukhopadhyay *et al.* (1995), and Shuai *et al.* (1997). Understanding the excited states of conjugated polymers is a challenge because it is this intermediate parameter regime that is applicable to conjugated polymers.

Also shown in Fig. 5.2 is the $1^3A_g^-$ state, which becomes the lowest charge-transfer triplet exciton in the strong coupling limit.

Having qualitatively described the behaviour of the low-lying excited states as a function of the interaction strengths, we now discuss the weak and strong coupling limits in more detail. The following two sections discuss these limits in a rather formal, mathematical sense. In Section 5.5, however, we introduce the valence bond method to present a qualitative, pictorial representation of the weak and strong coupling limits.

5.2   The weak-coupling limit

5.2.1   *Undimerized chains*

A field theoretical analysis of the extended Hubbard model with nearest (V_1) and next-nearest neighbour (V_2) Coulomb interactions in the weak-coupling limit yields a U(1) Thirring model (Essler *et al.* 2001). This model exhibits an explicit separation of the charge and spin degrees of freedom. The spin degrees of freedom are described by gapless bosonic excitations, namely pairs of spinons (or spin-density waves). The charge degrees of freedom, on the other hand, are described by a sine-Gordon model. For certain parameter ranges this model predicts bound electron-hole pairs (or excitons), whose spectrum is determined by the equation,

$$E_n = 2\Delta \sin\left(\frac{n\pi\xi}{2}\right). \tag{5.11}$$

Here, 2Δ is the charge gap and

$$\xi = \frac{\beta^2}{1 - \beta^2}, \tag{5.12}$$

where β depends on the parameters in the model. Bound states exist in the regime $0 < \beta < 1/\sqrt{2}$ and the number of bound states, N_{ex}, is determined by

$$N_{\text{ex}} = \text{integer part of} \left[\frac{1 - \beta^2}{\beta^2}\right]. \tag{5.13}$$

For the Hubbard model ($V_1 = 0$ and $V_2 = 0$) $\beta = 1$ and so there are no bound states. Long range Coulomb interactions decrease the value of β leading to one or more bound states. Although this is a weak-coupling theory, it also works reasonably well in the intermediate regime, defined by $U \sim 4t$, as shown by a comparison to a numerical calculation in Section 6.4.

5.2.2   *Dimerized chains*

In the weak-coupling limit a dimerized chain is a band insulator, with a filled valence band and an empty conduction band, as described in Section 3.4. The low-lying excitations are interband particle-hole transitions which bind to create Mott-Wannier excitons. An effective-particle model is developed in detail in the next chapter to describe this limit. We will see that the $1^1B_u^-$ and $1^3B_u^+$ excitons are the $n = 1$ singlet and triplet bound states, split by an exchange energy, while the $2^1A_g^+$ and $1^3A_g^-$ excitons are the degenerate $n = 2$ singlet and triplet bound states. Here, n is the principal quantum number of the hydrogen-like bound particle-hole pair described in Section 6.2. The $n = 2$ state is less strongly bound than the $n = 1$ state, so as U increases, and the charge gap opens, the excitation energies of the $2^1A_g^+$ and $1^3A_g^-$ states grows more quickly than that of the $1^1B_u^-$ and $1^3B_u^+$ states. This behaviour is shown in Figs 5.2 and 6.3.

5.3   The strong-coupling limit

In the strong coupling limit, $U \gg t$, a half-filled system is a Mott-Hubbard insulator, rather than a band insulator. The energies of the charge-transfer (or ionic) exciton states diverge strongly from the energies of the spin-density-wave (or covalent) states. The former are described by a high-energy spinless fermion model, introduced in Section 5.3.2, while the latter are described by a low-energy dimerized Heisenberg antiferromagnet, introduced in the following section.

5.3.1   *Low-energy dimerized Heisenberg antiferromagnet*

A standard canonical transformation can be performed on the Pariser-Parr-Pople model that has the effect of integrating out the high energy physics, leaving only the low energy spin dynamics.[23] The effective low-energy Hamiltonian is the dimerized Heisenberg antiferromagnet,

$$H = \sum_i J_i \mathbf{S}_i \cdot \mathbf{S}_{i+1}, \tag{5.14}$$

where

$$\mathbf{S}_i = \sum_{\rho\rho'} c_{i\rho}^\dagger \sigma_{\rho\rho'} c_{i\rho'}, \tag{5.15}$$

σ are the Pauli spin matrices,

$$J_i = \frac{4t^2(1 - 2\delta_i + \delta_i^2)}{U - V_1}, \tag{5.16}$$

and δ_i is the dimerization parameter for the ith bond.

For a fixed geometry consider a staggered dimerization, that is $\delta_i = \delta(-1)^i$. Then we can consider the *dimer limit*, defined by $0 \ll \delta \lesssim 1$ and the weakly-dimerized limit, defined by $0 \lesssim \delta \ll 1$.

5.3.1.1   *Dimer limit: $0 \ll \delta \lesssim 1$*   In this extreme limit the chain is composed of alternating 'strong' and 'weak' bonds, with a singlet dimer on each strong bond in the ground state, as illustrated in Fig. 5.3. A triplet, or magnon, excitation breaks one of these bonds, and costs an energy J ($\equiv 4t^2/(U - V_1)$). Formally, the magnon can be considered as two strongly bound spin-1/2 objects, known as spinons (Affleck 1997). The first singlet excitation corresponds to two broken bonds in an overall singlet and costs an energy $2J$. These are both illustrated in Fig. 5.3.

5.3.1.2   *Weakly dimerized limit: $0 \lesssim \delta \ll 1$*   As the dimerization becomes weaker the two spinons comprising the magnon become less confined, with their

---

[23]Numerous text books describe this procedure. See, for example, Fulde (1993).

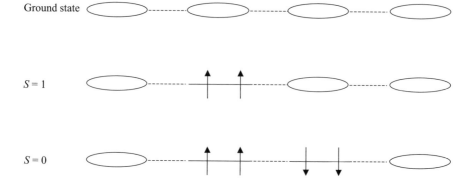

FIG. 5.3. A schematic representation of the ground state and the lowest triplet and singlet excitations of the dimerized quantum antiferromagnet. The ovals represent singlet dimers on the 'strong' bonds. This is an example of a valence bond representation. The valence bond method is introduced in Section 5.5.

separation scaling as $\sim \delta^{-2/3}$. The triplet ($1^3B_u^+$ state) excitation energy, $E(S = 1)$, vanishes as

$$E(S = 1) \sim \frac{J\delta^{2/3}}{\sqrt{|\ln(\delta)|}}. \tag{5.17}$$

For most of the parameter range the energy of the first singlet excitation (namely the $2^1A_g^+$ state), $E(S = 0)$, is twice the energy of the lowest triplet. However, as $\delta \to 0$ the singlet becomes a bound bimagnon, as

$$E(S = 0) \to \sqrt{3}E(S = 1) < 2E(S = 1), \tag{5.18}$$

(see Zheng *et al.* (2001) and references therein). Thus, only in the limit that $\delta \to 0$ is the $2^1A_g^+$ state a bound bimagnon.

5.3.2 *High-energy spinless fermion model*

In this section we derive an effective Hamiltonian that describes the high energy physics associated with particle-hole (or ionic) excitations across the charge gap. The Hamiltonian will describe a hole in the lower Hubbard band and a particle in the upper Hubbard band, interacting with an attractive potential. This attractive potential leads to bound, excitonic states. In the next chapter we derive an effective-particle model for these excitons. A real-space representation of an ionic state is illustrated in Fig. 5.5(b).

We write the Pariser-Parr-Pople model as

$$H = H^{\text{ke}} + H^{\text{pe}}, \tag{5.19}$$

where

$$H^{\text{ke}} = -t \sum_{i\sigma} (c_{i\sigma}^{\dagger} c_{i+1\sigma} + c_{i+1\sigma}^{\dagger} c_{i\sigma}) \tag{5.20}$$

and

$$H^{\text{pe}} = U \sum_{i} \left(N_{i\uparrow} - \frac{1}{2} \right) \left(N_{i\downarrow} - \frac{1}{2} \right) + \sum_{ij} V_j (N_i - 1)(N_{i+j} - 1). \tag{5.21}$$

It is instructive to recast H^{ke} as

$$H^{\text{ke}} = H_{\text{LHB}}^{\text{ke}} + H_{\text{UHB}}^{\text{ke}} + H_{\text{mix}}^{\text{ke}}, \tag{5.22}$$

where

$$H_{\text{LHB}}^{\text{ke}} = t \sum_{i\sigma} (1 - N_{i\bar{\sigma}})(c_{i\sigma}^{\dagger} c_{i+1\sigma} + c_{i+1\sigma}^{\dagger} c_{i\sigma})(1 - N_{i+1\bar{\sigma}}), \tag{5.23}$$

$$H_{\text{UHB}}^{\text{ke}} = t \sum_{i\sigma} N_{i\bar{\sigma}} (c_{i\sigma}^{\dagger} c_{i+1\sigma} + c_{i+1\sigma}^{\dagger} c_{i\sigma}) N_{i+1\bar{\sigma}}, \tag{5.24}$$

and

$$H_{\text{mix}}^{\text{ke}} = t \sum_{i\sigma} N_{i\bar{\sigma}} c_{i\sigma}^{\dagger} c_{i+1\sigma} (1 - N_{i+1\bar{\sigma}}) + N_{i+1\bar{\sigma}} c_{i+1\sigma}^{\dagger} c_{i\sigma} (1 - N_{i\bar{\sigma}})$$
$$+ \text{ hermitian conjugate.} \tag{5.25}$$

$H_{\text{LHB}}^{\text{ke}}$ describes the hopping of holes in the lower Hubbard band, $H_{\text{UHB}}^{\text{ke}}$ describes the hopping of double occupancies in the upper Hubbard band, while $H_{\text{mix}}^{\text{ke}}$ mixes the occupation of these two bands. By applying a canonical transformation to H, $H_{\text{mix}}^{\text{ke}}$ may be eliminated to $O(t/U)$ (Harris and Lange 1967). The elimination of $H_{\text{mix}}^{\text{ke}}$ implies that different occupations of the Hubbard bands are decoupled. The band width of each band is $4t$.

A particle-hole excitation from the ground state corresponds to creating a doubly occupied site, namely a particle, in the upper Hubbard band, and an empty site, namely a hole, in the lower Hubbard band. In the absence of long range Coulomb interactions described by H^{pe} the dynamics of the particle and hole are independently described by $H_{\text{UHB}}^{\text{ke}}$ and $H_{\text{LHB}}^{\text{ke}}$, respectively. The particle and hole move freely along the chain, irrespective of the underlying spin background. However, because of the elimination of $H_{\text{mix}}^{\text{ke}}$, the particle and hole cannot annihilate, so their positions cannot be exchanged. They therefore act as a pair of spinless fermions or hard core bosons. H^{pe} couples the particle and hole with an effective attractive interaction.

For this particle-hole excitation the N-body problem has thus been mapped onto the two-body problem, described by,

$$H_{\text{red}} = U - \sum_{ij} V_j (N_i - 1)(N_{i+j} - 1) - t \sum_{i} (a_i^{\dagger} a_{i+1} + a_{i+1}^{\dagger} a_i), \tag{5.26}$$

where $a_i^{\dagger}$ creates a spinless fermion on site i and $N_i = a_i^{\dagger} a_i$. For nearest neighbour interactions, $V_j = V_1 \delta_{1j}$, this two-body problem has an analytical solution, with

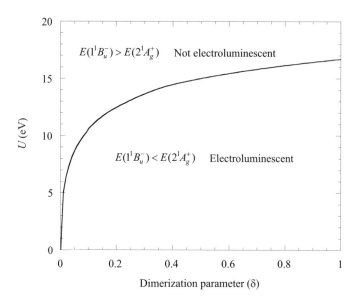

FIG. 5.4. The phase diagram of the Pariser-Parr-Pople model at half-filling. $t = 2.5$ eV.

the energy of the bound state being given by (Gallagher and Mazumdar 1997; Gebhard *et al.* 1997),

$$E(K) = U - V_1 - \frac{4t^2}{V_1} \cos^2\left(\frac{Ka}{2}\right), \qquad (5.27)$$

where K is the centre of mass momentum. Since the onset of the unbound particle-hole continuum is at

$$U - 4t \cos\left(\frac{K}{2}\right), \qquad (5.28)$$

a bound state only exists for $V_1 \geq 2|t|$. There is no analytical solution for a general $1/r$ interaction. However, a simple numerical solution is readily available by transforming the two-body problem into an effective particle problem, and in certain limits analytical results are also available. This is described in Section 6.3.

The minimum charge gap, 2Δ, is found by setting $K = 0$ in eqn (5.28), giving $2\Delta = U - 4t$, in agreement with eqn (5.7). Since the energies of the strong coupling excitons scale as 2Δ, we see that the large U behaviour is in agreement with the numerical results of Fig. 5.2.

5.4   The phase diagram of the undoped Pariser-Parr-Pople model

Having discussed the weak and strong coupling limits of the Pariser-Parr-Pople model, we can now qualitatively explain the behaviour of the excitation energies shown in Fig. 5.2.

In the weak-coupling limit the $2^1A_g^+$ exciton is less strongly bound than the $1^1B_u^-$ exciton. So, as the interaction strength increases, and the charge gap widens, the $2^1A_g^+$ energy initially increases faster than the $1^1B_u^-$ energy. However, as the $2^1A_g^+$ state acquires spin-density-wave character, its energy begins to decrease, so that there is a crossover in energy between the $1^1B_u^-$ and $2^1A_g^+$ states. This crossover is a function of δ and U/t, and is shown in a phase diagram in Fig. 5.4. Generally, for a smaller δ or a larger U/t, the $2^1A_g^+$ state becomes more correlated, and lies below the $1^1B_u^-$ state (Mukhopadhyay et al. 1995). This crossover has important consequences for the electroluminescent properties of conjugated polymers. As we show in Chapter 8, the $1^1B_u^-$ state is dipole connected to the ground state, whereas the $2^1A_g^+$ state is not. Thus, if $E(1^1B_u^-) < E(2^1A_g^+)$ the polymer is electroluminescent, whereas if $E(2^1A_g^+) < E(1^1B_u^-)$ it is not, and the system decays nonradiatively to the ground state.

5.5   The valence bond method

We now introduce the valence bond method which provides a pictorial representation of the weak and strong coupling limits.

The Bloch or molecular orbital states are exact eigenstates of the kinetic energy operator. We therefore might expect these states to be a useful basis in the weak-coupling limit where the kinetic energy dominates over the potential energy. Conversely, in the strong-coupling limit, where the potential energy dominates over the kinetic energy, we might expect that a real-space basis is more appropriate. The valence bond method provides such a real-space basis. In this section we give a brief description of the valence bond method, as it provides insight into the crossover from weak to strong coupling. Furthermore, as we explain in Section 7.3, it also provides insight into the effects of electronic interactions on the strength of the bond alternation. This method is also used in Appendix G to qualitatively explain the lowest-lying singlet excitation of benzene. The reader is referred to Coulson (1961), Mazumdar and Soos (1979), or Baeriswyl et al. (1992) for more details of the valence bond method.

At half-filling the real-space basis states can be characterized by the number of doubly occupied sites (with the same number of empty sites). Basis states with no doubly occupied sites are classed as 'covalent', whereas basis states with one or more doubly occupied site are classed as 'ionic'. In a covalent basis state each site is linked to one other by a singlet bond.

We can illustrate this point most simply with a two-π orbital system (e.g. ethylene or a dimer). In this system the singlet subspace is spanned by three basis states illustrated in Fig. 5.5: the covalent basis state (a) and the two equivalent basis states in (b). The triplet state is also illustrated. (Notice that Fig. 5.3 also illustrates covalent basis states.)

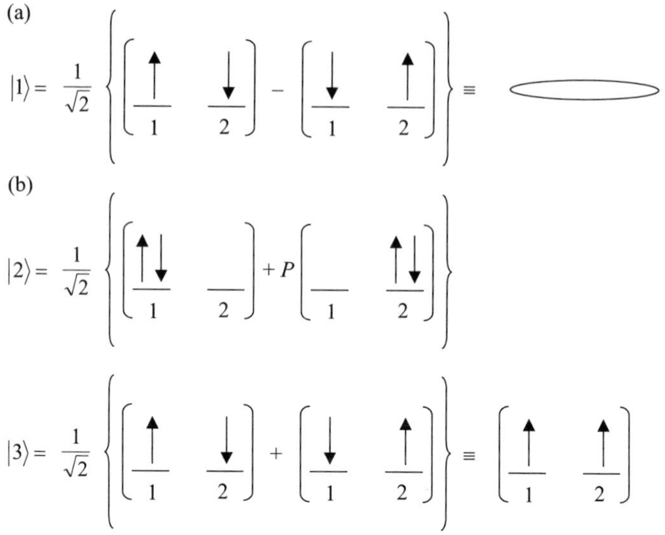

FIG. 5.5. Valence bond basis states of the two-site dimer. $|1\rangle$ and $|2\rangle$ are singlets, while $|3\rangle$ is the triplet, showing the $S_z = 0$ and $S_z = 1$ representations. The parity eigenvalue $P = \pm 1$.

The basis states are formally represented as,

$$|1\rangle = \frac{1}{\sqrt{2}} \left(c_{1\uparrow}^\dagger c_{2\downarrow}^\dagger - c_{1\downarrow}^\dagger c_{2\uparrow}^\dagger \right) |0\rangle, \qquad (5.29)$$

$$|2\rangle = \frac{1}{\sqrt{2}} \left(c_{1\uparrow}^\dagger c_{2\downarrow}^\dagger + P c_{2\uparrow}^\dagger c_{2\downarrow}^\dagger \right) |0\rangle, \qquad (5.30)$$

and

$$|3\rangle = \frac{1}{\sqrt{2}} \left(c_{1\uparrow}^\dagger c_{2\downarrow}^\dagger + c_{1\downarrow}^\dagger c_{2\uparrow}^\dagger \right) |0\rangle, \qquad (5.31)$$

where the parity eigenvalue $P = \pm 1$.

Using these basis states the singlet eigenstates of the dimer may be expressed as,

$$|\psi_1\rangle = a|1\rangle + b|2\rangle \qquad (5.32)$$

with $P = 1$,

$$|\psi_2\rangle = |2\rangle \qquad (5.33)$$

with $P = -1$, and

$$|\psi_3\rangle = b|1\rangle - a|2\rangle \qquad (5.34)$$

with $P = 1$.

Table 5.1 *The molecular orbital eigenstates of the dimer (namely the noninteracting limit of the Pariser-Parr-Pople model) expressed within the valence bond basis*

State	Energy	b/a	State label
$|\psi_1\rangle$	$-2t$	1	$^1A_g^+$
$|\psi_4\rangle$	0	—	$^3B_u^+$
$|\psi_2\rangle$	0	—	$^1B_u^-$
$|\psi_3\rangle$	$2t$	1	$^1A_g^+$

Table 5.2 *The eigenstates of the dimer in the strong-coupling limit of the Pariser-Parr-Pole model ($J = 4t^2/(U - V_1)$ and V_1 is the nearest neighbour Coulomb repulsion)*

State	Energy	b/a	State label
$|\psi_1\rangle$	$V_1 - J$	$2t/(U - V_1)$	$^1A_g^+$
$|\psi_4\rangle$	V_1	—	$^3B_u^+$
$|\psi_2\rangle$	U	—	$^1B_u^-$
$|\psi_3\rangle$	$U + J$	$2t/(U - V_1)$	$^1A_g^+$

The triplet state is

$$|\psi_4\rangle = |3\rangle. \tag{5.35}$$

We first describe the noninteracting solutions. These are the molecular orbital eigenstates listed in Table 5.1. The goundstate is a linear superposition of the covalent and ionic basis states, $|1\rangle$ and $|2\rangle$. The first excited singlet state is the odd-parity ionic state, $|2\rangle$, whereas the triplet excitation is the covalent state $|3\rangle$. The second singlet excitation is an antisymmetric linear combination of $|1\rangle$ and $|2\rangle$.

In the strong-coupling limit (defined by $(U - V_1) >> t$) the eigenstates evolve smoothly to those listed in Table 5.2. Now the ground state is predominately the covalent state $|1\rangle$. The covalent triplet state has an excitation energy J. The first singlet excitation is again the ionic state $|2\rangle$ with an excitation energy $(U - V_1 + J)$,[24] while the final singlet is again predominately ionic with an excitation energy $(U - V_1 + 2J)$.

We therefore see in the dimer example that in the strong-coupling limit the spectrum has split into low-energy covalent states, with an energy scale set by J, and high-energy ionic states with an energy scale set by $U - V_1$. This simple picture essentially confirms the discussions of Section 5.3, except for three caveats arising from there being only two sites and two electrons. First, the ionic spectrum is split-off from the covalent spectrum by $U - V_1$ rather than by U for widely separated singly and doubly occupied sites. Second, the even parity singlet excitation is not related to the 2^1A_g state of large systems in the strong-coupling limit. Finally, the triplet state on a dimer has no ionic character. As

---

[24]The energy difference between the $^1B_u^-$ and $^3B_u^+$ states, $U - V_1$, (which is valid for all nonzero interactions) is precisely the singlet-triplet exchange energy derived in the Mott-Wannier exciton limit described in Section 6.2 and Appendix D.

discussed earlier this is not representative of larger systems, where the $1^3B_u^+$ state evolves from a Mott-Wannier exciton at weak-coupling to a gapped spin density wave at large coupling.

In the next chapter we focus entirely on bound particle-hole, or excitonic, excitations.

6

EXCITONS IN CONJUGATED POLYMERS

6.1 Introduction

The study of excitons in conjugated polymers has often been inspired by the treatment of excitons in bulk three-dimensional semiconductors (as described in Knox (1963)). A particle-hole excitation from the valence band to the conduction band in a semiconductor leaves a positively charged hole in the valence band and a negatively charged electron in the conduction band. The Coulomb attraction between these particles results in bound states, or excitons. In three-dimensional semiconductors the excitons are usually weakly bound, with large particle-hole separations, and are well described by a hydrogenic model. Excitons in this limit are known as *Mott-Wannier* excitons.

This model of bound conduction band electrons and valence band holes can also be applied to conjugated polymers (Abe *et al.* 1992, Abe 1993). In conjugated polymers a one-dimensional hydrogenic model applies. However, a difference between one and three dimensions is that in one-dimension the first excited state (namely the lowest bound state) is generally strongly bound, with a small particle-hole separation (Loudon 1959). Such strongly bound excitons are akin to *Frenkel* excitons in molecular crystals, which are delocalized intra-atomic excitations. Molecular crystals also exhibit *charge-transfer* excitons, defined as excitons with larger particle-hole separations, so that the exciton wavefunction is spread over a few molecules. Finally, in molecular crystals, the term Mott-Wannier exciton is usually reserved for excitons with a very large particle-hole separation.

For simplicity, however, we prefer to denote all excitons formed from bound states of conduction band electrons and valence band holes as Mott-Wannier excitons, recognizing that this term includes both small and large radius excitons. We call this limit the *weak-coupling* limit, as the starting point in the construction of the exciton basis is the noninteracting band limit. As we will see, a real space description of a Mott-Wannier exciton is of a hole in a valence band Wannier orbital bound to an electron in a conduction band Wannier orbital.

An opposite, *strong-coupling* limit has also been used to describe excitons in conjugated polymers (Gallagher and Mazumdar 1997; Gebhard *et al.* 1997; Essler *et al.* 2001; Barford 2002). As described in the previous chapter, in this limit a correlation gap separates the electron removal spectral weight (the lower Hubbard band) from the electron addition spectral weight (the upper Hubbard band). Now the bound particle-hole excitations are *Mott-Hubbard* excitons. That is, a particle excited from the lower Hubbard band to the upper Hubbard band is bound to the hole it leaves behind. In a real-space picture this corresponds to two electrons in the same atomic orbital bound to an empty atomic orbital moving in a sea of singly occupied orbitals. A one-dimensional hydrogenic model also applies in this limit (Barford 2002).

Generally, conjugated polymers are in the intermediate regime, as the band width is comparable to the interaction strength. In other words, the electronic kinetic energy is comparable to the electronic potential energy, and so neither the weak nor strong coupling limits apply. Since no theory has yet been developed for the intermediate regime, numerical calculations are the only means to theoretically study this regime.

In this chapter we describe the theory of excitons in isolated conjugated polymers. We start with the weak-coupling limit and describe Mott-Wannier excitons. Next, we discuss the strong coupling limit and Mott-Hubbard excitons. Finally we discuss the intermediate coupling regime. The weak to strong coupling crossover described in this chapter is also discussed by Mazumdar and Chandross (1997). In the weak and strong coupling limits we derive relatively simple *effective-particle* models to describe the physics of excitons. These effective-particle models are the prototypes for more sophisticated approaches that are better at quantitatively predicting excited state energies. These are described briefly in Section 6.2.4. The utility of the simpler approach presented here is that it gives qualitative and intuitive insight into the physics of excitons in conjugated polymers.

6.2   The weak-coupling limit

The weak-coupling limit takes as its starting point the conventional semiconductor noninteracting band picture, introduced in Chapter 3.[25] The ground state is an occupied valence-band and an empty conduction-band. A bound conduction band electron and valence band hole move through the lattice as an effective-particle. In this section we derive the effective-particle model, discuss its solutions and compare them to essentially exact calculations on the same Hamiltonian (Barford *et al.* 2002b). We develop this theory for a linear, dimerized chain.

6.2.1   *The effective-particle model*

Since excitons are bound particle-hole excitations, a convenient basis for their description are the particle-hole basis states introduced in Chapter 3. In k-space these basis states are $\{|k_e, k_h\rangle\}$, defined by

$$|k_e, k_h\rangle = \frac{1}{\sqrt{2}} \left(c_{k_e\uparrow}^{c\dagger} c_{k_h\uparrow}^{v} \pm c_{k_e\downarrow}^{c\dagger} c_{k_h\downarrow}^{v} \right) |GS\rangle, \qquad (6.1)$$

where the plus sign creates a singlet basis and the minus sign creates a triplet basis. The ground state, $|GS\rangle$, is the filled Fermi sea and is defined in eqn (3.28). The basis state $|k_e, k_h\rangle$ is a function of the centre-of-mass momentum,

[25]The reader may find it helpful to review Chapter 3 before reading this section.

$$K = (k_e - k_h),$$ (6.2)

and the relative momentum,

$$2k' = (k_e + k_h).$$ (6.3)

Thus,

$$|k_e, k_h\rangle \equiv |k' + K/2, k' - K/2\rangle.$$ (6.4)

This particle-hole excitation is illustrated in Fig. 3.4. Now, for translationally invariant Hamiltonians K is a good quantum number. However, unlike the non-interacting Hamiltonian, the interacting Hamiltonian mixes states with different k'.

The general exciton eigenstate, $|\Phi_K^{MW}\rangle$, is therefore a linear superposition of these basis states

$$|\Phi_K^{MW}\rangle = \sum_{k'} \Phi(k', K)|k' + K/2, k' - K/2\rangle.$$ (6.5)

As shown in Section 3.6.1, as a consequence of particle-hole symmetry, the amplitudes satisfy

$$\Phi(k', K) = \pm\Phi(-k', K),$$ (6.6)

if the particle-hole eigenvalue, $J = \mp 1$ for singlet excitations, or $J = \pm 1$ for triplet excitations.

To proceed further we need an equation for $\Phi(k', K)$, which may be obtained from an effective exciton Hamiltonian. However, as the Coulomb interaction is not diagonal in k-space, a real space basis leads to a more intuitive description.[26]

A basis state in real-space, introduced in Section 3.6.1, is

$$|R + r/2, R - r/2\rangle = S_{rR}^\dagger |GS\rangle,$$ (6.7)

where the operator $S_{rR}^\dagger$ creates a particle-hole excitation from the ground state, $|GS\rangle$. These are defined as,

$$S_{rR}^\dagger = \frac{1}{\sqrt{2}} \left(c_{R+r/2,\uparrow}^{c\dagger} c_{R-r/2,\uparrow}^{v} \pm c_{R+r/2,\downarrow}^{c\dagger} c_{R-r/2,\downarrow}^{v} \right)$$ (6.8)

and

$$|GS\rangle = \prod_R c_{R\uparrow}^{v\dagger} c_{R\downarrow}^{v\dagger} |0\rangle,$$ (6.9)

respectively. $c_{R\sigma}^{v\dagger}$ are the Wannier operators, defined in Section 3.4.1.1. As in the k-space picture, the plus sign in eqn (6.8) creates a singlet basis and the minus sign creates a triplet basis. This basis state is shown schematically in Fig. 6.1.

---

[26]The reader is referred to the work of Abe *et al.* (1992), for an analysis in k-space.

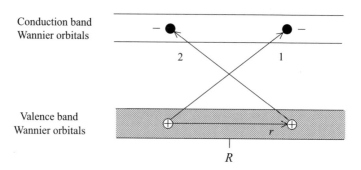

FIG. 6.1. The real-space particle-hole excitation, $|R+r/2, R-r/2\rangle$, labelled 1, from the valence band Wannier orbital at $R - r/2$ to the conduction band valence orbital at $R+r/2$. Its degenerate counterpart, $|R-r/2, R+r/2\rangle$, connected by the particle-hole transformation, is labelled 2. $R = (r_e + r_h)/2$ is the centre-of-mass coordinate and $r = (r_e - r_h)$ is the relative coordinate. A Mott-Wannier exciton is a bound particle-hole pair in this representation.

$$R = \frac{(r_e + r_h)}{2} \tag{6.10}$$

is the centre-of-mass coordinate and

$$r = (r_e - r_h) \tag{6.11}$$

is the relative coordinate. R and r are discrete variables measured as a contour length along the polymer chain. Thus, defining d as the contour length between repeat units (e.g. $2a$ for a dimerized chain), r/d is the number of repeat units between the electron and hole.

We now define the general exciton eigenstate, $|\Phi^{\mathrm{MW}}\rangle$, as

$$|\Phi^{\mathrm{MW}}\rangle = \sum_{r,R} \Phi(r, R)|R + r/2, R - r/2\rangle, \tag{6.12}$$

where $\Phi(r, R)$ is the exciton wave function, which is obtained from the appropriate exciton Hamiltonian. Since the exciton is a two-particle bound state, we can proceed to find solutions in analogy to the hydrogen atom. Thus, we introduce the effective-particle model by separating the centre-of-mass and relative coordinates. For periodic boundary conditions we assume that

$$\Phi_{nK}(r, R) = \psi_n(r)\Psi_K(R), \tag{6.13}$$

where $\Psi_K(R)$ is the centre-of-mass wavefunction,

$$\Psi_K(R) = \frac{1}{\sqrt{N_u}} \exp(iKR), \tag{6.14}$$

and K is the centre-of-mass momentum: $-\pi/d \leq K \leq \pi/d$.

$\psi_n(r)$ is the relative wavefunction that describes the internal structure of the exciton. Owing to particle-hole symmetry it satisfies

$$\psi_n(r) = \pm\psi_n(-r), \quad (6.15)$$

if the particle-hole eigenvalue, $J = \mp 1$ for singlet excitations, or $J = \pm 1$ for triplet excitations.[27]

For open boundary conditions we assume that

$$\Phi_{nj}(r, R) = \psi_n(r)\Psi_j(R), \quad (6.16)$$

where $\Psi_j(R)$ is the centre-of-mass wavefunction,

$$\Psi_j(R) = \sqrt{\frac{2}{N_u + 1}} \sin(\beta_j R), \quad (6.17)$$

and β_j is the centre-of-mass pseudomomentum:

$$\beta_j = \frac{j\pi}{(N_u + 1)d}, \quad (6.18)$$

and $j = 1, 2, \cdots, N_u$.

As shown in Appendix D, the relative wavefunction, $\psi_n(r)$, satisfies the following Schrödinger difference equation,

$$-2\tilde{t}\cos\left(\frac{Kd}{2}\right)(\psi_n(r+d) + \psi_n(r-d)) + \left(2X\delta_{r0}\delta_M - \tilde{V}(r)\right)\psi_n(r)$$
$$= \left(E - \tilde{U} - 2\Delta + X\right)\psi_n(r), \quad (6.19)$$

where $\delta_M = 1$ for singlet excitons, and $\delta_M = 0$ for triplet excitons. $\delta_{r0} = 1$ when $r = 0$ and $\delta_{r0} = 0$ when $r \neq 0$. E is the energy of the effective-particle. For effective-particles on linear chains K is replaced by β_j.

The model parameters are defined as

- The effective hybridization integral, $\tilde{t} = t(1 - \delta)/2$.
- The HOMO-LUMO gap, $2\Delta = 2t(1 + \delta)$.
- The local electron-hole interaction, $\tilde{U} = (U + V_1)/2$.
- The long-range electron-hole interaction, $\tilde{V}(r)$.
- The singlet-triplet exchange interaction, $2X = U - V_1$.[28]

t and δ are parameters from the Pariser-Parr-Pople model (defined in eqn (2.52)) and V_j is the Ohno potential (defined in eqn (2.55)). U therefore sets the scale of the electronic interactions. As $r \to \infty$ the electron-hole potential

$$\tilde{V}(r) \to V(\epsilon_{\text{eff}}r), \quad (6.20)$$

where $V(r)$ is the Ohno potential and ϵ_{eff} is the effective dielectric constant arising from the polymer geometry, as explained in Appendix D.

---

[27] We note that $\psi(r)$ is the Fourier transform of $\Phi(k')$ with respect to k'.

[28] See Appendix D for a discussion of the origin of this term.

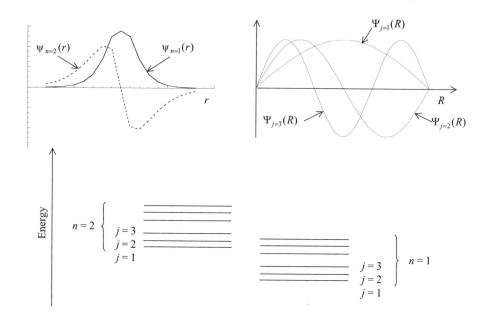

FIG. 6.2. The effective-particle model of excitons on a linear chain. The total exciton wavefunction, $\Phi_{nj}(r, R) = \psi_n(r)\Psi_j(R)$, where $\psi_n(r)$ is the relative wavefunction and $\Psi_j(R)$ is the centre-of-mass wavefunction. For each principal quantum number, n, there is a family of excitons with different pseudomomentum, β_j, which form a band of exciton states. The relative wavefunctions are solutions of eqn (6.19), while the centre-of-mass wavefunctions are determined by eqn (6.17).

Notice that two quantum numbers specify the exciton eigenstates, eqn (6.13) or eqn (6.16): the principle quantum number, n, and the (pseudo) momentum quantum number, K (or β_j). For every n there are a family of excitons with different centre-of-mass momenta, and hence different centre-of-mass kinetic energy. Odd and even values of n correspond to the relative wavefunction, $\psi_n(r)$, being even or odd under a reversal of the relative coordinate, respectively. We refer to even and odd parity excitons as excitons whose relative wavefunction is even or odd under a reversal of the relative coordinate. This does not mean that the overall parity of the eigenstate (eqn (6.12)), determined by both the centre-of-mass and relative wavefuctions, is even or odd. The number of nodes in the exciton wavefunction, $\psi_n(r)$, is $n-1$. Figure 6.2 illustrates the wavefunctions and energies of excitons in the effective-particle model.

There is an important observation to be made about this effective-particle model. This is that since the exchange interaction, X, is local (i.e. it is only nonzero when $r = 0$), we immediately see that this term vanishes for odd parity excitons (namely, $\psi_n(r) = -\psi_n(-r)$), as $\psi_n(0) = 0$. Now, since the parity of the exciton is determined by the particle-hole symmetry, and odd singlet and triplet excitons are determined by positive and negative particle-hole symmetries, respectively, this theory predicts that $^1A_g^+$ and $^3A_g^-$, and $^1B_u^+$ and $^3B_u^-$ excitons are degenerate.[29]

Eqn (6.19) is the Schrödinger equation for describing the Mott-Wannier exciton wavefunctions and energies. In more sophisticated treatments it is usually known as the Bethe-Salpeter equation. In the following two sections the solutions of this equation will be described.

6.2.2 Solutions of the effective-particle model

The continuum or effective-mass limit and the hydrogenic solutions of eqn (6.19) are described in Appendix E. As the continuum limit analysis remains qualitatively correct for a discrete lattice, it is useful to summarize them here before discussing the general solutions.

- Odd parity (even n) states follow the Rydberg series, defined by the Rydberg number $n' = n/2$. Thus, the binding energies are

$$E_n(K) = \frac{E_I(K)}{(n/2)^2}, \qquad (6.21)$$

  where E_I is the effective Rydberg, defined in eqn (E.10).
- The lowest even parity state ($n = 1$) is strongly bound, with a binding energy scaling as $V(r \to 0) \to \infty$.
- The remaining even parity (odd n) states are bounded in energy by the odd parity states.

The solutions on a discrete lattice are illustrated in Figs 6.3 and 6.4. Figure 6.3 shows the excitation energies of the two lowest singlet and triplet states, and the charge gap, while Fig. 6.4 shows the exciton probability density, $\psi_n^2(r)$.

6.2.3 Comparisons to the numerical calculations

It is instructive to compare the approximate weak-coupling theory to essential exact, numerical (density matrix renormalization group) calculations on the same model (namely the Pariser-Parr-Pople model). The numerical calculations are performed on polymer chains with the polyacetylene geometry. Since these chains posses inversion symmetry the many-body eigenstates are either even (A_g) or odd (B_u). As discussed previously, the singlet exciton wavefunction has either even or odd parity when the particle-hole eigenvalue is odd or even. Conversely, the triplet exciton wavefunction has either even or odd parity when the particle-hole eigenvalue is even or odd. As a consequence, we can express a $^1B_u^-$ state as

$$|^1B_u^-\rangle = \sum_{\text{odd } n} \sum_{\text{odd } j} A_{nj} |\Phi_{nj}^{\text{MW}}\rangle + \text{other contributions}, \qquad (6.22)$$

---

[29]This statement is only approximately correct, as the exchange term is formally only local for onsite Coulomb interactions, that is, when $V_j = 0$. However, nonlocal exchange interactions are negligible and decay rapidly: the exchange parameter between two dimers ℓ units apart is $X_\ell = (2V_{2\ell} - V_{2\ell-1} - V_{2\ell+1})/4 \approx -1/\ell^3$ for $1/r$ Coulomb interactions.

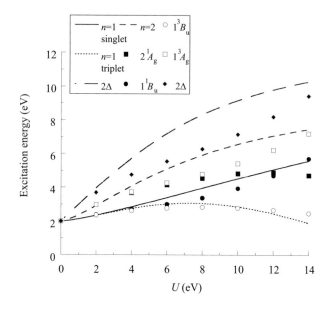

FIG. 6.3. The transition energies of the lowest momentum ($j = 1$) $n = 1$ singlet (solid curve), $n = 1$ triplet (dotted curve), $n = 2$ singlet and triplet (short-dashed curve) excitons, and the charge gap (long-dashed curve) in the weak-coupling limit. $t = 2.5$ eV and $\delta = 0.2$. The symbols are the DMRG calculations of the Pariser–Parr–Pople model on 102 site chains: filled circles, $1^1B_u^-$ ($n = 1, j = 1$ singlet); full circles, $1^3B_u^+$ ($n = 1, j = 1$ triplet); filled squares $2^1A_g^+$ ($n = 2, j = 1$ singlet); open squares, $1^3A_g^-$ ($n = 2, j = 1$ triplet); diamonds, charge gap.

where $|\Phi_{nj}^{\mathrm{MW}}\rangle$ is defined by eqns (6.12) and (6.16). Similarly, we can express a $^1A_g^+$ state as

$$|^1A_g^+\rangle = \sum_{\text{even } n}\sum_{\text{odd } j} A_{nj}|\Phi_{nj}^{\mathrm{MW}}\rangle + \text{other contributions}. \qquad (6.23)$$

Generally, the sums will be dominated by one component (except at anticrossings, as discussed shortly). The 'other contributions' to the state vectors include components not described by the exciton basis, for example, covalent and holon-doublon terms. These are expected to be negligible in the weak-coupling limit.

Table 6.1 classifies the many-body states according to their exciton quantum numbers.[30]

---

[30]Notice that odd n (i.e. even parity $\psi(r)$) and odd j (i.e. even centre-of-mass wavefunction) implies $^1B_u^-$ and even n (i.e. odd parity $\psi(r)$) and odd j implies $^1A_g^+$, as the reflection operator reflects both the centre-of-mass and relative coordinates, and hence exchanges the electron and hole.

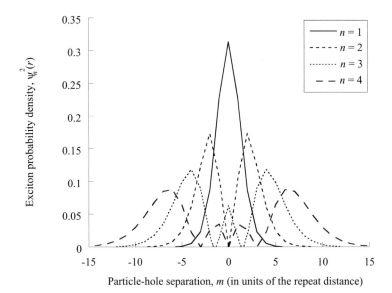

FIG. 6.4. The singlet exciton probability density, $\psi_n^2(r)$, in the weak-coupling limit. $t = 2.5$ eV, $U = 3.33$ eV and $\delta = 0.2$.

Table 6.1 *The classification of the many body singlet exciton states with particle-hole symmetry in terms of their Mott-Wannier exciton quantum numbers (The correspond-ing triplet states with the same spatial symmetry but opposite particle-hole symmetry have the same quantum numbers)*

	Exciton quantum numbers	
State	n	j
$^1A_g^+$	Even	Odd
$^1B_u^+$	Even	Even
$^1B_u^-$	Odd	Odd
$^1A_g^-$	Odd	Even

6.2.3.1 *Transition energies* The evolution of the calculated exciton energies as function of chain length shows a number interesting features. Fig. 6.5(a) and (b) shows the $^1B_u^-$ (odd n, odd j) and $^1A_g^+$ (even n, odd j) spectra, respectively, for representative weak-coupling parameters. The different pseudomomentum (j) states for the same n, and anticrossings between states of different n are clearly seen. Figure 6.5(a) shows the $n = 1$ and $n = 3$ excitons converging to 2.6 eV and 3.9 eV, respectively, while Fig. 6.5(b) shows the $n = 2$ and $n = 4$ excitons converging to 3.5 eV and 4.1 eV, respectively. The band gap is also shown converging to 4.4 eV. Thus, for 102 sites with these parameters, there are at least four families of bound excitons.

For large N the energies scale as $1/N^2$. A detailed analysis (Barford *et al.*

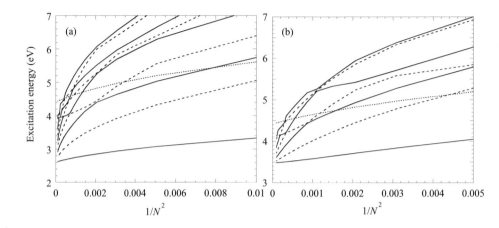

FIG. 6.5. The DMRG calculated singlet exciton transition energies of the Pariser-Parr-Pople model as a function of square of the inverse chain length. The parameters are representative of the weak-coupling limit: $t = 2.5$ eV, $U = 3.33$ eV, and $\delta = 0.2$. All curves are for odd pseudomomentum quantum number, j. Solid and dashed curves are to illustrate the anticrossings. Also shown is the charge gap as the dotted curve. (a) $^1B_u^-$ states, showing the $n = 1$ exciton converge to 2.6 eV and the $n = 3$ exciton converge to 3.9 eV. (b) $^1A_g^+$ states, showing the $n = 2$ exciton converge to 3.5 eV and the $n = 4$ exciton converge to 4.1 eV. Notice that the $^1B_u^-$ states are interleaved with $^1A_g^-$ states (odd n and even j), while the $^1A_g^+$ states are interleaved with $^1B_u^+$ states (even n and even j). These states are not shown.

2002b) shows that the ratios of the gradients of the energies versus $1/N^2$ of the three lowest pseudomomentum branches (namely $j = 1, 3$, and 5) of the $^1B_u^-$ ($n = 1$) and $^1A_g^+$ ($n = 2$) excitons scale as $1 : 9 : 25$. This agrees with the analysis of Appendix E, indicating particle-in-a-box behaviour for the effective-particle. (See eqn (E.7) with K replaced by $\beta_j = j\pi/(N_u + 1)d$.) For short chain lengths, however, the $1/N^2$ scaling is replaced by a $1/N$ scaling. This crossover reflects a break-down of the effective-particle model, which occurs when the chain length becomes comparable to or less than the particle-hole separation. In this limit the excitation energies increase as the chain length decreases, because of a confinement effect associated with squeezing the electron and hole together.

6.2.3.2 *Particle-hole correlation function* The exciton component of the numerical many-body wavefunction can be obtained using the operator $S_{rR}^\dagger$ defined in eqn (6.8). Using eqns (6.7) and (6.12) it follows that projecting $S_{rR}^\dagger|GS\rangle$ onto the exciton state $|\Phi_{nj}^{MW}\rangle$ gives the exciton wavefunction $\Phi_{nj}(r, R)$:

$$\Phi_{nj}(r, R) = \langle\Phi_{nj}^{MW}|S_{rR}^\dagger|GS\rangle. \tag{6.24}$$

Thus, for a general numerically evaluated eigenstate, $|p\rangle$,

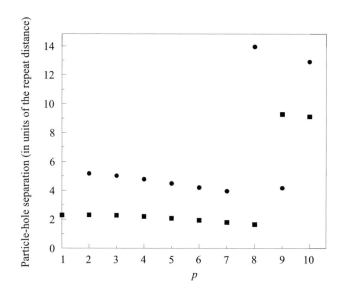

FIG. 6.6. The DMRG calculated root-mean-square particle-hole separations, r_p (eqn (6.26)) in units of the molecular repeat distance, for 102 sites. $t = 2.5$ eV, $U = 3.33$ eV, and $\delta = 0.2$. $p^1B_u^-$ states (squares) and $p^1A_g^+$ states (circles). The molecular repeat distance is twice the lattice distance.

$$\Phi_p(r, R) \equiv \sum_{nj} A_{nj}^p \Phi_{nj}(r, R) \approx \langle p | S_{rR}^\dagger | \text{true GS} \rangle, \qquad (6.25)$$

where $|\text{true GS}\rangle$ is the numerically evaluated ground state. $\Phi_p(R, r)$ is the overall exciton wavefunction for the state $|p\rangle$.

Correlation functions can now be calculated from this wavefunction. For example, the average particle-hole separation, r_p, is (Barford et al. 1998),

$$r_p^2 = \frac{\sum_{r,R} r^2 \Phi_p^2(r, R)}{\sum_{r,R} \Phi_p^2(r, R)}. \qquad (6.26)$$

The particle-hole separations are shown in Fig. 6.6 at 102 sites. The jumps in the separation occur at $p = 9$ and $p = 8$ for the even and odd parity excitons, respectively, corresponding to the $j = 1$ branches of the $n = 3$ and $n = 4$ families of excitons. Notice that, as predicted in Appendix E, the particle-hole separations decrease with increasing j for the same n.

6.2.3.3 *Exciton families* To compare the numerical results with the weak-coupling exciton model we need to identify the exciton families (labelled by n) and their pseudomomentum branches (labelled by j). There at least three ways of identifying the lowest pseudomomentum branch ($j = 1$) of a given exciton family.

Table 6.2 *The excitation energies and binding energies (in eV) for the first four $j = 1$ Mott-Wannier singlet excitons of a 102 site chain ($\delta = 0.2$, $U = 3.33$ eV, and $t = 2.5$ eV)*

State	DMRG calculation		Weak-coupling theory	
	Excitation energy	Binding energy	Excitation energy	Binding energy
$1^1B_u^-$ ($n = 1$)	2.62	1.82	2.68	2.49
$2^1A_g^+$ ($n = 2$)	3.49	0.95	3.70	1.47
$9^1B_u^-$ ($n = 3$)	3.93	0.51	4.25	0.92
$8^1A_g^+$ ($n = 4$)	4.13	0.31	4.54	0.63

- These states have strong dipole moments between different families if $|n - n'|$ is odd (and thus they contribute strongly to the nonlinear optical spectroscopies, as we describe in Chapter 8).
- There are sharp changes in the particle-hole separations, r_p, as shown in Fig. 6.6.
- As shown in Fig. 6.5, energy plots against inverse chain length identify the different exciton families by the bands of states which converge as the chain length increases.

The comparisons between the essentially exact calculations and the weak-coupling theory are summarized by comparing their predictions in Table 6.2 and Fig. 6.3. For small U the agreement between the excitation energies is good for the $n = 1$ and $n = 2$ excitons, and reasonably good for the $n = 3$ and $n = 4$ excitons. In this limit, as predicted, the odd parity singlet and triplet excitons are degenerate. However, for intermediate U the results are less good, particularly for the odd parity excitons. The binding energies do not agree well. The origin of this disagreement is that the unbound particle-hole pair is strongly solvated by intrachain screening, which is absent in the simple theory presented here. The excitons are also screened, but this screening becomes less strong as the excitons become more strongly bound.[31]

6.2.3.4 *Primary excitons* We conclude this section with a few remarks on the *essential states* responsible for the nonlinear optical susceptibilities. As described in Chapter 8, there are at most four states in a particular excitation pathway in the sum-over-states calculation of the third-order nonlinear susceptibility, $\chi^{(3)}$. Only a few excitation pathways (and hence states) contribute to this sum. The pathway must contain strong dipole moments to the ground state. In the weak coupling limit these are the $1^1A_g^+$, $1^1B_u^-$, $2^1A_g^+$, and $n^1B_u^-$ states, namely the ground state and the $j = 1$, $n = 1, 2$, and 3 Mott-Wannier excitons.[32]

---

[31] In practice, the intrachain excitations are also screened by interchain interactions, and thus the DMRG calculations also overestimate the experimentally determined binding energies by ca. 1 eV. Interchain screening is discussed in Chapter 9.

[32] The anticrossings between a higher j of a lower n with the $j = 1$ state of a higher n, shown in Fig. 6.5, can lead to spurious 'essential states', as oscillator strength is transferred from the $j = 1$ state of higher n to the higher j state of the lower n. These other essential states, arising

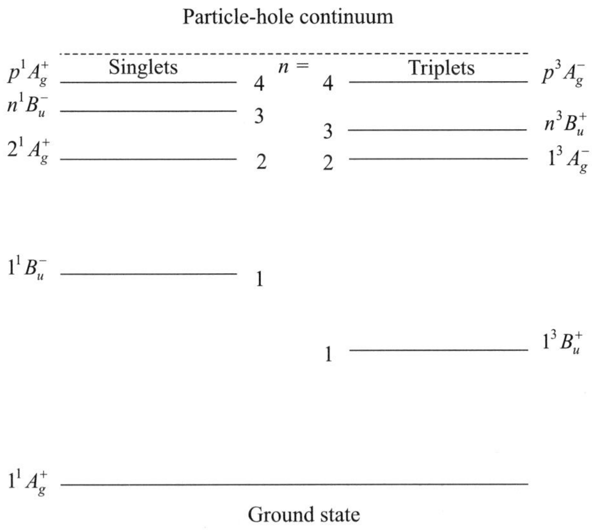

FIG. 6.7. A schematic energy level diagram of the $j = 1$ members of the Mott-Wannier exciton families. The symmetry assignments refer to centro-symmetric polymers with particle-hole symmetry.

Figure 6.7 is a schematic energy level diagram that summarizes the key low-lying states with their predominant Mott-Wannier exciton quantum number assignments. The symmetry assignments refer to centro-symmetric polymers with particle-hole symmetry. Notice, however, that particle-hole symmetry is not an exact symmetry of conjugated polymers. The symmetry assignments in the absence of particle-hole symmetry are shown in Table 6.3. In particular notice that the $2^1 A_g^+$ state becomes the $m^1 A_g$ state (where $m > 2$), as this state is not in general the lowest excited even-parity singlet state. The lowest excited even-parity singlet state is the $1^1 A_g^-$ state, which is the $j = 2$ pseudomomentum state associated with the $n = 1$ exciton.

6.2.4 Refinements of the theory

A severe approximation in the weak-coupling effective model presented in this chapter is that the ground state is noninteracting, and thus there is no screening of the electron-hole interactions by the other π-electrons. Such screening is usually modelled by a static dielectric constant in the electron-hole interaction (eqn (6.20)), and often also by a renormalization of the charge gap, 2Δ.

Obviously, Hartree-Fock or density functional theory (DFT) ground states would be more rigorous approximations. With such starting points, most meth-

from the accidental degeneracies, are quite different from the competing essential states seen in the intermediate-coupling regime, as discussed below.

Table 6.3 *State labels and corresponding exciton quantum numbers in the weak-coupling limit*

State labels		Exciton quantum numbers	
Particle-hole symmetry	No particle-hole symmetry	n	j
$1^1A_g^+$	1^1A_g	–	–
$1^1B_u^-$	1^1B_u	1	1
$1^1A_g^-$	2^1A_g	1	2
$2^1A_g^+$	m^1A_g	2	1
$n^1B_u^-$	n^1B_u	3	1
$1^3B_u^+$	1^3B_u	1	1
$1^3A_g^+$	1^3A_g	1	2
$1^3A_g^-$	m^3A_g	2	1
$n^3B_u^+$	n^3B_u	3	1

ods proceed in a similar spirit to that presented here. Namely, a basis of single particle-hole excitations is constructed from the ground state and the Hamiltonian is diagonalized within this basis. The single configuration interaction (S-CI) method proceeds precisely in this manner (see (Szabdo and Ostlund 1996) or (Atkins and Friedman 1997)). The random phase approximation improves on the S-CI method by constructing a polarizable Hartree-Fock ground state. Finally, the GWA-Bethe-Salpeter equation method takes as its starting point a DFT-local density approximation (LDA) ground state, which is then corrected by the GW-approximation before a Bethe-Salpeter equation is constructed for the electron-hole wavefunction using the LDA orbitals (Rohfling and Louie 1999; van der Horst *et al.* 2000, 2001).

While the methods described here give quantitative predictions of exciton energies and wave functions in the weak-coupling limit, they are inapplicable for the strong-coupling limit, described in the next section, as the excited state basis of conduction band particles and valence band holes is no longer valid.

6.3 The strong-coupling limit

Excitons in the strong-coupling limit are quite different from their counterparts in the weak-coupling limit. In the weak-coupling limit excitons are particle-hole excitations from the valence to the conduction band. As described in Section 6.2, a real space picture corresponds to a particle in a local antibonding molecular orbital bound to a hole in local bonding molecular orbital. Since an electron and hole can exist on the same dimer there are no restrictions on the symmetries of the relative wavefunction, and both singlet and triplet excitons exist. However, as shown in Section 5.3.2, the strong-coupling limit starts from the approximation that the Coulomb interactions are so large that the undimerized band splits into a lower and upper Hubbard band. At half-filling the lower Hubbard band is full, corresponding to one electron per π orbital. Now an exciton is a particle in the upper Hubbard band bound to a hole in the lower Hubbard band. An equivalent,

(a)

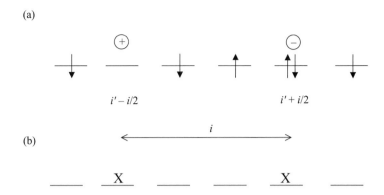

(b)

FIG. 6.8. A Mott-Hubbard exciton. (a) An empty orbital (holon) at $i' - i/2$ is bound to a doubly occupied orbital (doublon) at $i' + i/2$. This is equivalent to two spinless fermions, or hardcore bosons, represented by Xs, shown in (b).

real-space picture is of an empty orbital bound to a doubly occupied orbital on another site, as illustrated in Fig. 6.8. These are Mott-Hubbard excitons. This problem maps onto the problem of two bound spinless fermions (or hard core bosons), as described in Section 5.3.2. The particle and hole cannot exist on the same site, so there is a local hardcore repulsion, and the relative wavefunction is zero for $r = 0$. As will be shown in the next section, in the continuum limit with a $1/r$ potential the bound states form a Rydberg series, with each energy level being composed of an even and odd pair of states (Barford 2002).

6.3.1 The effective-particle model

The general particle-hole eigenstate in this limit is of the form

$$|\Phi^{MH}\rangle = \sum_{ii'} \Phi_{ii'} |i' + i/2, i' - i/2\rangle, \qquad (6.27)$$

where

$$|i' + i/2, i' - i/2\rangle = a^\dagger_{i'+i/2} a^\dagger_{i'-i/2} |0\rangle, \qquad (6.28)$$

and the MH refers to Mott-Hubbard excitons. $a^\dagger_i$ creates a spinless fermion on site i and $|0\rangle$ is the vacuum of the two-body problem.

Following the same procedure as in Section 6.2.1, using H_{red} (defined in eqns (5.26) and (6.27)), the relative wavefunction $\psi_n(i)$, satisfies

$$-2t \cos\left(\frac{Ka}{2}\right)(\psi_n(i+1) + \psi_n(i-1)) - V(i)\psi_n(i) = (E - U)\psi_n(i), \qquad (6.29)$$

where i is the distance between atomic the pair of spinless fermions and a is the lattice spacing, as illustrated in Fig. 6.8. $V(i)$ is the Coulomb potential between a pair of electrons i sites apart. Equation (6.29) is the Schrödinger equation for describing Mott-Hubbard excitons.

The hard core repulsion, imposed by the condition $\psi_n(0) = 0$, implies that even and odd parity solutions are degenerate, because $\psi_n(i)$ can be matched by either $\pm\psi_n(-i)$ at the origin. Thus, the solutions of eqn (6.29) in the continuum limit are precisely the hydrogen atom wavefunctions (for zero angular momentum), as described in Appendix E. In practice, the degeneracy between the even and odd parity solutions is lifted by virtual transitions between the Hubbard bands. To second order in perturbation theory the energy splitting is $O(t^2/(U - V_1))$.

In analogy with eqns (6.22) and (6.23) we can express the exciton states as

$$|^1B_u^-\rangle = \sum_{\text{odd } n \text{ odd } j} \sum B_{nj}|\Phi_{nj}^{\text{MH}}\rangle + \text{other contributions,} \qquad (6.30)$$

and

$$|^1A_g^+\rangle = \sum_{\text{even } n \text{ odd } j} \sum B_{nj}|\Phi_{nj}^{\text{MH}}\rangle + \text{other contributions,} \qquad (6.31)$$

where $|\Phi_{nj}^{\text{MH}}\rangle$ is defined in eqn (6.27).

Since the unbound continuum starts at $(U - 4t)$, we see that this model is unphysical for $U \lesssim 4t$, as then the bound states would have a negative excitation energy. So, although we can obtain binding energies, we cannot obtain physically realistic excitation energies in the intermediate coupling regime. However, as we shall see in the next section, this theory does provide qualitative insight to the behaviour of the intermediate-coupling regime.

This strong-coupling exciton theory completely neglects the low-lying spin density wave excitations (described in Chapter 5); nor does it describe the triplet excitons. In this limit the $1^3B_u^+$ state has evolved from the $n = 1$, $j = 1$ Mott-Wannier triplet exciton to a gapless spin-density-wave (or magnon), while the $2^1A_g^+$ state has evolved from the weak-coupling $n = 2$, $j = 1$ Mott-Wannier exciton to a pair of triplets (or a bimagnon) (Schulten and Karplus 1972; Tavan and Schulten 1987). This picture is confirmed by the numerical calculations for six sites, presented in Table 6.4. The first odd parity singlet exciton is now the $m^1A_g^+$ state, where $m > 2$ ($m = 5$ for the six-site calculation). This, as predicted, is virtually degenerate with its associated even parity exciton, the $1^1B_u^-$ state, being ca. $4t^2/(U - V_1)$ higher in energy. The $1^3A_g^-$ state is the $1^3B_u^+$ triplet bound to the $1^1B_u^-$ exciton, while the $8^3B_u^+$ state is the $1^3B_u^+$ triplet bound to the $5^1A_g^+$ exciton. The excited states in this limit are represented schematically in Fig. 6.9.

6.3.1.1 *The particle-hole correlation function*   In analogy to the weak-coupling limit, the strong coupling theory suggests an appropriate particle-hole (or holon-doublon) correlation function for measuring the particle-hole separation in numerical calculations. Suppose that

Table 6.4 *Excitation energies (in eV) of the key low-lying states for the undimerized six-site chain (U = 100 eV, t = 2.5 eV, and V(i) is determined by the Ohno potential, eqn (2.55))*

State	Character	Excitation Energy (eV)
$2^1 A_g^+$	Pair of magnons	0.365
$1^1 B_u^-$	$n = 1, j = 1$ Mott-Hubbard singlet exciton	86.375
$5^1 A_g^+ \ (\equiv m^1 A_g^+)$	$n = 2, j = 1$ Mott-Hubbard singlet exciton	86.650
$n^1 B_u^-$	$n = 3, j = 1$ Mott-Hubbard singlet exciton	—
$1^3 B_u^+$	Magnon	0.138
$1^3 A_g^-$	Magnon bound to the $n = 1$ Mott-Hubbard exciton	86.545
$8^3 B_u^+$	Magnon bound to the $n = 2$ Mott-Hubbard exciton	86.819

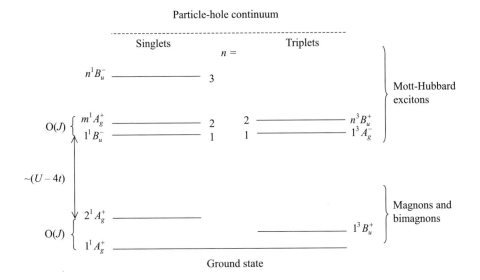

FIG. 6.9. Schematic energy level diagram of the spin-density-wave states and the $j = 1$ members of the Mott-Hubbard exciton families. $J = 4t^2/(U - V_1)$. The symmetry assignments refer to centro-symmetric polymers with particle-hole symmetry.

$$h_i^\dagger = \sum_\sigma c_{i\sigma}(1 - n_{i\bar\sigma}) \qquad (6.32)$$

creates a holon (namely removes a particle from the lower Hubbard band), while

$$d_i^\dagger = \sum_\sigma c_{i\sigma}^\dagger n_{i\bar\sigma} \qquad (6.33)$$

creates a doublon (namely a particle in the upper Hubbard band). Then, assuming that the ground state, $|GS\rangle$, is constructed by occupying the lower Hubbard band, the wavefunction for a holon-doublon excitation separated by $r = ia$ in the excited state $|p\rangle$ is

$$\Phi_p(r, R) = \langle p|d^\dagger_{i'+i/2}h^\dagger_{i'-i/2}|GS\rangle. \tag{6.34}$$

The mean-square separation is then given by eqn (6.26).

6.4 The intermediate-coupling regime

As the strength of the Coulomb interactions are increased from the weak-coupling limit the character of the ground state and excitations changes. As discussed already, a new class of excitations emerges, and these are the spin-density-wave (or covalent) states. The lowest lying triplet $(1^3B_u^+)$ becomes a spin-density-wave, and the $2^1A_g^+$ state is a bimagnon. A higher lying $^1A_g^+$ state evolves into the $n = 2$ Mott-Hubbard exciton.

The intermediate-coupling regime is in the crossover between these limits. In fact, the crossover also occurs as a function of the dimerization, δ. Consider the undimerized chain, with $\delta = 0$. As a result of the perfect nesting in one-dimension there is always a correlation gap in the electronic spectrum of the half-filled chain for any nonzero Coulomb interaction. For the Hubbard model the correlation gap is $\sim \sqrt{Ut}\exp(-t/U)$ for $t \gg U$, while it is $(U - 4t)$ for $t \ll U$. We expect this prediction of a gapped charge spectrum to remain correct for long-range interactions. The correlation gap separates the lower and upper Hubbard bands. A particle-hole excitation across the correlation gap will result in a bound Mott-Hubbard exciton for any interaction strength, although for weak interactions the exciton will be considerably more complicated than the holon-doublon exciton described in the last section. Alternatively, if the dimerization gap $(2\Delta = 4\delta t)$ is large compared to the correlation gap, we expect Mott-Wannier excitons to be the dominant low-energy particle-hole excitations.

We can see this behaviour by studying the numerical calculations. First, consider $\delta = 0$. Figure 6.10(a) shows the four lowest essential states. The $1^1B_u^-$, $9^1A_g^+$, and $7^1B_u^-$ states are the $j = 1$, $n = 1, 2$, and 3 Mott-Hubbard excitons. The $2^1A_g^+$ state, with an energy lower than the $1^1B_u^-$ state, is predominately a bimagnon. The particle-hole separations in the holon-doublon channel, r_p, defined by eqns (6.26) and (6.34) are also shown.

Next, consider $\delta = 0.2$. Fig. 6.10(b) shows that the four lowest essential states appear to fit the weak-coupling model, as they are the $1^1B_u^-$, $2^1A_g^+$, and $4^1B_u^-$ states. These are $j = 1$, $n = 1, 2$, and 3 Mott-Wannier excitons.

At $\delta = 0.1$ there are both Mott-Hubbard and Mott-Wannier excitons, forming two inter-related families of essential states. In general, the $^1B_u^-$ states are linear superpositions of eqns (6.22) and (6.30), while the $^1A_g^+$ states are linear superpositions of eqns (6.23) and (6.31). As the bond dimerization decreases the spin-density-wave component of the $2^1A_g^+$ state increases (Mukhopadhyay et al. 1995). Figure 6.10(c) shows the $1^1B_u^-$, $2^1A_g^+$, and $4^1B_u^-$ states, predominately forming the Mott-Wannier family of excitons, while Fig. 6.10(d) shows the $1^1B_u^-$,

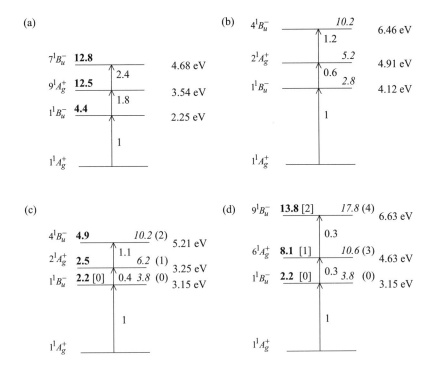

FIG. 6.10. The DMRG calculated essential states (defined as the four lowest states with the strongest interstate dipole moments) for 30 site chains. The arrows show the transition dipole moments normalized to $1^1A_g^+ \rightarrow 1^1B_u^-$ transition dipole moment. The Mott-Wannier exciton and Mott-Hubbard exciton particle-hole separations (in units of the lattice spacing) are shown italicized and bold, respectively. (a) $t = 2.5$ eV, $U = 10$ eV, and $\delta = 0$, showing the Mott-Hubbard series; (b) $t = 2.5$ eV, $U = 10$ eV, and $\delta = 0.2$, showing the Mott-Wannier series; (c) $t = 2.5$ eV, $U = 10$ eV, and $\delta = 0.1$; and (d) $t = 2.5$ eV, $U = 10$ eV, and $\delta = 0.1$. Also shown in (c) and (d) are the number of nodes in the Mott-Wannier exciton wavefunction (in round brackets ( )) and the number of nodes in the Mott-Hubbard exciton wavefunction (in square brackets [ ]). In all cases long range interactions are determined by the Ohno potential. (The results for the undimerized chain (a) agree reasonably well with the weak-coupling field theory described in Section 5.2.1 (Essler *et al.* 2001). Using eqns (5.11) and (5.13) with the charge gap $2\Delta = 4.0$ eV and $E_1 = 2.25$ eV implies that there are two bound states and that $E_2 = 3.72$ eV, compared to the calculated value of 3.54 eV.)

$6^1A_g^+$, and $9^1B_u^-$ states, predominately forming the Mott-Hubbard family of excitons. The progression of excitons in both families can also be identified by the jumps in the relevant particle-hole separation.

Since the $1^1B_u^-$ state has large dipole moments to both the $2^1A_g^+$ and $6^1A_g^+$ states, this state clearly has large amplitudes in both the $n = 1$ Mott-Hubbard and $n = 1$ Mott-Wannier families. Similarly, the character of each state can be investigated by examining their Mott-Wannier and Mott-Hubbard exciton wavefunction components (defined by eqns and 6.25 and 6.34, respectively). The number of nodes in the Mott-Wannier and Mott-Hubbard exciton wavefunctions are shown in Fig. 6.10(c) and (d). We note that the $6^1 A_g^+$ state has one node in the Mott-Hubbard exciton wavefunction and three nodes in the Mott-Wannier exciton wavefunction. Thus, this state is an admixture of the $n = 2$ Mott-Hubbard exciton and the $n = 4$ Mott-Wannier exciton. Since this lies energetically below the $4^1 B_u^-$ state, which is predominately the $n = 3$ Mott-Wannier exciton, we see that the simple classification of essential states into Mott-Wannier or Mott-Hubbard excitons fails in certain parameter regimes.

6.5   Concluding remarks

In this chapter we have described the effective-particle models for excitons in the weak and strong coupling limits, and compared them to essential exact, numerical (DMRG) calculations. We saw that there is good agreement between the effective-particle models and the computational results in these limits. We used these extreme limits to understand the numerical calculations in the intermediate-coupling regime. We summarize the main points as follows:

- In the weak-coupling limit (where the single particle gap is larger than the correlation gap) the bound states are Mott-Wannier excitons, namely conduction band electrons bound to valence band holes. A Mott-Wannier exciton in real space is an electron in a conduction band Wannier orbital bound to a hole in a valence band Wannier orbital. Singlet and triplet excitons whose relative wavefunctions are odd under a reflection of the relative coordinate (namely, even n excitons) are degenerate. Thus, the $2^1 A_g^+$ and $1^3 A_g^-$ states are degenerate in this limit. In contrast, singlet and triplet excitons whose relative wavefunctions are even under a reflection of the relative coordinate (namely, odd n excitons) have energies that are split by the exchange interaction.

- In the strong-coupling limit (where the correlation gap is larger than the single particle gap) the bound states are Mott-Hubbard excitons, namely particles in the upper Hubbard band bound to holes in the lower Hubbard band. A Mott-Hubbard exciton in real space is a doubly occupied atomic orbital bound to an empty atomic orbital. These bound states occur in doublets of even and odd parity excitons. Triplet excitons are magnons bound to the singlet excitons, and hence are degenerate with their singlet counterparts.

- In the intermediate-coupling regime Mott-Wannier excitons are the more appropriate description for large dimerization ($\delta = 0.2$), while for the undimerized chain Mott-Hubbard excitons are the correct description. For dimerizations relevant to polyacetylene and polydiacetylene (that is, $\delta \sim$

0.1) there is a mixed representation of both Mott-Hubbard and Mott-Wannier excitons.

- For both weak and strong coupling an infinite number of bound states exist for $1/r$ interactions for an infinite polymer. As a result of the discreteness of the lattice, and the restrictions on the exciton wavefunctions in one-dimension, the progression of states does not follow the Rydberg series.

- Formally, the exciton binding energy is defined relative to the energy of a widely separated uncorrelated electron-hole pair. In practice, excitons whose particle-hole separation exceeds the length of the polymer (or more correctly, the conjugation length) can be considered unbound. This marks the breakdown of the effective-particle model.

- It is not known how many bound states exist in the intermediate regime.

- The numerical calculations show that the $n = 1$ exciton binding energy increases monotonically with increasing Coulomb interaction. At large coupling the binding energy agrees with the strong-coupling theory. We may therefore place a theoretical estimate on the binding energy of excitons in isolated conjugated polymers as ca. 4.6 eV.

- The numerically calculated exciton excitation energies scale as the inverse of the chain length for short chains, and the inverse of the square of the chain length for long chains. The long chain limit reflects the particle-in-a-box behaviour of the effective-particle, where the energy decreases as the chain length increases because of the delocalization of the effective particle. However, when the chain length is comparable to or shorter than the particle-hole separation we expect the effective-particle model to break down. In that limit the excitation energies increase with decreasing chain length because of confinement effects associated with squeezing the electron and hole together.

- The so-called $m^1 A_g$ state observed in nonlinear optical spectroscopy (as described in Chapter 8) is the $2^1 A_g^+$ state, or $n = 2$ Mott-Wannier exciton in the weak-coupling limit. In the strong-coupling limit it is the $m^1 A_g^+$ state, or $n = 2$ Mott-Hubbard exciton. This is often referred to as the charge-transfer exciton, owing to its larger electron-hole separation in comparison to the more strongly bound $1^1 B_u^-$ state, or $n = 1$ exciton.

It is instructive to apply these exciton theories to actual conjugated polymers. Calculations on single poly($para$-phenylene) chains (see Section 11.2.3) predict the $1^1 B_{1u}^-$ ($n = 1$, $j = 1$) exciton at 3.7 eV, the $2^1 A_g^+$ ($n = 2$, $j = 1$) exciton at 5.1 eV and the $1^3 A_g^-$ triplet close in energy to the $2^1 A_g^+$ state, at 5.5 eV. This progression indicates a Mott-Wannier series of excitons. An equivalent description applies to poly($para$-phenylene-vinylene). In contrast, polyacetylene and polydiacetylene have predominately Mott-Hubbard excitons. In polyacetylene the vertical energies of the $1^1 B_{1u}^-$ and $2^1 A_g^+$ states are virtually degenerate (see Section 10.2.1), while for polydiacteylene the $2^1 A_g^+$ state lies a few tenths of an eV higher than the $1^1 B_{1u}^-$ state (Race et $al.$ 2001). In both cases the $^1 A_g^+$

state most strongly connected to the $1^1B_{1u}^-$ state is not the $2^1A_g^+$ state, but a higher $m^1A_g^+$ state, fitting the pattern of Mott-Hubbard excitons. Furthermore, in both cases the $2^1A_g^+$ state undergoes strong electron-lattice relaxation, and its relaxed energy lies below that of the relaxed $1^1B_{1u}^-$ state (Race $et\ al.$ 2003). This places polyacetylene and polydiacetylene on the correlated side of the intermediate-coupling regime.

These predictions apply to the vertical excitations of single polymer chains. Various additional intrinsic and extrinsic effects can significantly modify excited state energies. Covalent states, such as the highly correlated $1^3B_u^+$ and $2^1A_g^+$ states, undergo significant electron-lattice relaxation, and as already stated, this leads to a reversal of the $1^1B_{1u}^-$ and $2^1A_g^+$ energies in polyacetylene and polydiacetylene. In the next chapter we describe the combined effects of electron-electron and electron-phonon coupling. There it will be shown that the description of the triplet excited state as a bound particle-hole pair - even in the weak-coupling limit - requires revision.

An important extrinsic effect is screening by the environment, and again, this significantly alters the energy of excited states. States with larger binding energy are less screened than those that are weakly bound. Current estimates are that the $n = 1$ exciton solvates by ca. 0.3 eV, the $n = 2$ exciton solvates by ca. 0.6 eV, and the band gap solvates by ca. 1.5 - 2.0 eV (Moore and Yaron 1998). This effect will be discussed in Chapter 9.

Finally, we remark that this chapter has focussed on excitons confined to single chains. However, the formalism can easily be extended to interacting chains, and this will be considered briefly in Section 9.6 when we describe a theory for the singlet exciton yield in light emitting polymers.

7

ELECTRON-LATTICE COUPLING II: INTERACTING ELECTRONS

7.1 Introduction

It has been recognized for some 30 years that neither electron-phonon inter-actions nor electron-electron interactions alone are capable of explaining the electronic properties of conjugated polymers. In *trans*-polyacetylene, for exam-ple, it is impossible to consistently predict the values of the bond alternation and the optical gap within a noninteracting framework for reasonable values of the electron-phonon coupling constant (Ovchinnikov *et al.* 1973). In fact, as de-scribed in this chapter, the bond alternation depends crucially on the strength of the electronic interactions (Horsch 1981), with the optical gap being significantly enhanced.

The effects of electron-phonon interactions alone were described in Chapter 4. We showed that these interactions lead to a dimerized, semiconducting ground state and to solitonic structures in the excited states. On the other hand, the effects of electron-electron interactions in a polymer with a fixed geometry were described in Chapters 5 and 6. There it was shown that the electronic interactions cause a metal-insulator (or Mott-Hubbard) transition in undimerized chains. Electron-electron interactions also cause Mott-Wannier excitons in the weak-coupling limit of dimerized chains, and to both Mott-Hubbard excitons and spin density wave excitations in the strong coupling limit.

In this chapter we describe the combined effects of both electron-electron and electron-phonon interactions, focussing our attention on these effects in linear polyenes. As well as significantly enhancing the bond alternation in the ground state, we will see that these combined effects lead to a rich and complex behaviour in the excited states.

We start this investigation by treating the electronic degrees of freedom within the Born-Oppenheimer approximation, where the nuclear degrees of free-dom are static, classical variables. The π-electron model that describes both electron-electron and electron-phonon interactions in the Born-Oppenheimer ap-proximation is known as the Pariser-Parr-Pople-Peierls model. This is described and its predictions are analyzed in the following sections. Chapter 10 will deal with quantum phonons in an interacting electron model, specifically for *trans*-polyacetylene.

7.2   The Pariser-Parr-Pople-Peierls model

The Pariser-Parr-Pople-Peierls (P-P-P-P) model, H_{PPPP}, is defined as the Peierls model (defined in Section 4.2) supplemented by the Coulomb interactions. The electrons and lattice are coupled together by the effects of changes in the bond lengths both on the one-electron transfer integrals and the Coulomb interactions. These effects are generally treated up to first order in the changes of bond length. As the density-density correlator, $(N_m - 1)(N_n - 1)$, decays rapidly with distance, it is also a reasonable approximation to retain changes in the Coulomb potential for only nearest neighbour interactions.

Thus, the Pariser-Parr-Pople-Peierls model is defined as

$$H_{\text{PPPP}} = -2\sum_n t_n \hat{T}_n + W \sum_n \Delta_n (N_{n+1} - 1)(N_n - 1) \tag{7.1}$$

$$+ \frac{1}{4\pi t\lambda}\sum_n \Delta_n^2 + \Gamma \sum_n \Delta_n + U\sum_n \left(N_{n\uparrow} - \frac{1}{2}\right)\left(N_{n\downarrow} - \frac{1}{2}\right)$$

$$+ \frac{1}{2}\sum_{m\neq n} V_{mn}(N_m - 1)(N_n - 1),$$

where

$$\hat{T}_n = \frac{1}{2}\sum_\sigma \left(c^\dagger_{n+1,\sigma}c_{n,\sigma} + c^\dagger_{n,\sigma}c_{n+1,\sigma}\right), \tag{7.2}$$

$$t_n = t + t_e(-1)^n + \frac{\Delta_n}{2}, \tag{7.3}$$

$$\Delta_n = -2\alpha(u_{n+1} - u_n), \tag{7.4}$$

and

$$\lambda = \frac{2\alpha^2}{\pi K t}. \tag{7.5}$$

The variables and parameters in eqns (7.2) - (7.5) are defined in Chapter 4. V_{mn} (defined in eqn (2.55)) is the Ohno potential for the undistorted structure. For generality, we also include an extrinsic dimerization, represented by the term $t_e(-1)^n$ in eqn (7.3).

The second term on the right-hand side of eqn (7.1) is the change in the Coulomb interactions arising from changes in bond length, where

$$W = \frac{1}{2\alpha}\left(\frac{\partial V_{mn}}{\partial \mathbf{r}_{mn}}\right)_{\mathbf{r}_{mn}=r_0} = \frac{U\gamma r_0}{2\alpha(1 + \gamma r_0^2)^{3/2}}, \tag{7.6}$$

$\gamma = (U/14.397)^2$ and r_0 is the undistorted average bond length. It is instructive to rewrite this term as

$$-2\alpha W \sum_n (u_{n+1} - u_n)(N_{n+1} - 1)(N_n - 1), \tag{7.7}$$

where we have used eqn (7.4). Expanding and resumming we see that this term has two components. One component is the electron-phonon coupling arising from the change in the ionic potentials,

$$-2\alpha W \sum_n (u_{n+1} - u_{n-1}) N_n. \tag{7.8}$$

Comparing this with eqn (2.40) we identify $2\alpha W$ with β. The other component represents the changes in the nearest neighbour electron-electron interaction from the change in bond length,

$$-2\alpha W \sum_n (u_{n+1} - u_n) N_{n+1} N_n. \tag{7.9}$$

The first term on the right-hand side of eqn (7.1) is just the electron-phonon coupling arising from the change in the kinetic energy, which is the first term in eqn (2.40).

As described in Chapter 4, by using the Hellmann-Feynman theorem we can derive a self-consistent equation for $\{\Delta_n\}$ for any state,

$$\Delta_n = 2\pi t \lambda \left(\langle \hat{T}_n \rangle - W \langle \hat{D}_n \rangle - \Gamma \right), \tag{7.10}$$

where

$$\hat{D}_n = (N_{n+1} - 1)(N_n - 1) \tag{7.11}$$

is the density-density correlator for the nth bond.

We note that constant chain lengths, implying that $\sum_n \Delta_n = 0$, means that

$$\Gamma = \frac{1}{N} \sum_n \left(\langle \hat{T}_n \rangle - W \langle \hat{D}_n \rangle \right) = \overline{\langle \hat{T}_n \rangle} - W \overline{\langle \hat{D}_n \rangle}, \tag{7.12}$$

where the overbar represents the spatial average. Using the definition of Γ from eqn (4.8), eqn (7.12) derives eqn (2.46). We now see from eqn (7.10) that the distortion of the nth bond is proportional to the deviation of $\langle \hat{T}_n \rangle - W \langle \hat{D}_n \rangle$ from its average value, Γ. Thus, both the bond order, $\hat{T}_n$, and the bond density-density correlator, $\hat{D}_n$, contribute to this distortion.

7.3 Dimerization and optical gaps

In this section we describe how electron-electron interactions modify the noninteracting predictions both for the bond alternation of the ground state and the optical gap. This subject has been studied by a number of authors (Ukranskii 1978; Horsch 1981; Hirsch 1983; Mazumdar and Dixit 1983; Baeriswyl and Maki 1985; Konig and Stollhoff 1990; Baeriswyl et al. 1992).

The ground state staggered dimerization, Δ_0, is defined as

$$\Delta_0 = \frac{1}{N} \sum_n \Delta_n (-1)^n. \tag{7.13}$$

Figure 7.1 shows the behaviour of Δ_0 for a linear chain as a function of the Coulomb interaction strength, U. As U increases to the intermediate values of

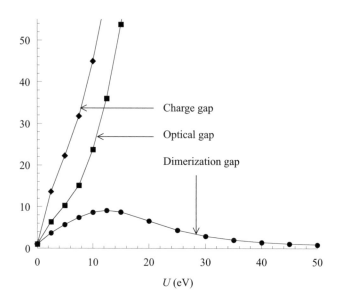

FIG. 7.1. The ground state dimerization, Δ_0, (circles), the optical gap, defined as
    the vertical excitation energy of the lowest dipole-allowed singlet (the $1^1B_u^-$ state)
    (squares) and the vertical charge gap (diamonds). All values are extrapolations to
    the infinite chain and are normalized to their $U = 0$ values. In the $U = 0$ limit
    the optical and charge gaps equal $2\Delta_0$. $t = 2.5$ eV and $\lambda = 0.1$. These results were
    calculated from the Pariser-Parr-Pople-Peierls model on linear polyenes with the
    trans-polyacetylene geometry with $r_0 = 1.4$ Å and $\alpha = 4.6$ eVÅ^{-2}.

$10 - 15$ eV we see that there is a substantial increase in the bond alternation,
being approximately ten times greater than the noninteracting value. As the
Coulomb interaction increases further, however, the dimerization decreases and
vanishes in the asymptotic limit. The dimerization is maximized at $U \sim 4t$, a
result first qualitatively explained by Dixit and Mazumdar (1984).

To understand the origins of this behaviour we first need to consider the
valence bond diagrams that dominate the wavefunction associated with a particular geometrical structure. As described in Section 5.5, a valence bond diagram
is a real-space representation of the electronic basis states. Figure 7.2 shows the
A and B geometrical structures of a dimerized linear chain. Also shown are the A
and B Kekulé valence bond diagrams that dominate the ground state wavefunction for each geometrical structure. The A and B phases are equivalent under a
translation.

The ground state of the A-phase geometrical structure will be dominated by
the A-phase Kekulé diagram. However, there will also be a contribution from
the B-phase Kekulé diagram because of quantum fluctuations (and vice versa).
These fluctuations will reduce the magnitude of the dimerization. Thus, the

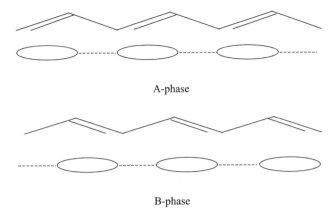

FIG. 7.2. The dimerized geometrical structures and associated Kekulé valence bond diagrams of linear polyenes.

dimerization is enhanced if quantum fluctuations are suppressed. In the weak coupling limit this suppression is achieved by electronic interactions as they increase the resonance barrier between equivalent valence bond diagrams.

To see why interactions increase the resonance barrier between equivalent diagrams consider Fig. 7.3. This figure shows the two equivalent Kekulé valence bond diagrams of benzene linked together by different valence bond diagrams generated by the action of kinetic energy operator. In these diagrams a line represents a singlet bond, a cross represents a doubly occupied site and a dot represents an empty site. A diagram composed entirely of singly occupied sites is termed covalent, whereas a diagram containing one or more occupied sites is termed ionic.

As shown by Coulson and Dixon (1961) the resonance barrier between two equivalent diagrams increases with the lengths of the paths and decreases with the number of paths connecting these diagrams. Now, electronic interactions make ionic configurations energetically less favourable because each doubly occupied site costs an energy U. Thus, interactions also effectively reduce the number of paths and therefore increase the resonance barrier, thereby enhancing the dimerization.

This argument breaks down when charge fluctuations are suppressed in favour of spin-density wave fluctuations. As described in Chapter 5 this occurs in the intermediate-coupling regime, around $U \sim 4t$.

We can see why the dimerization *decreases* for strong electronic interactions by considering the Pariser-Parr-Pople-Peierls model in the strong-coupling limit (defined by $U \gg t$). As described in Section 5.3.1, in this limit the low-energy physics of the Pariser-Parr-Pople model is described by the Heisenberg antiferromagnetic. Similarly, the low-energy physics of the Pariser-Parr-Pople-Peierls model is described by the Heisenberg-Peierls (or spin-Peierls) model,

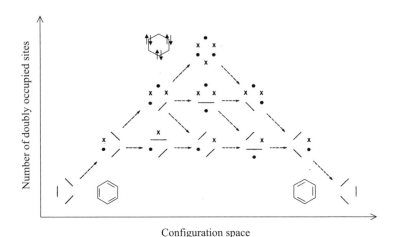

Configuration space

FIG. 7.3. Some representative paths connecting the two Kekulé diagrams of benzene (shown in the bottom left and right corners). The diagram in the top corner shows the fully ionic configuration. All other diagrams are shown without their carbon atom vertices. A line represents a singlet bond, a cross ($\times$) represents a doubly occupied site and a dot ($\cdot$) represents an empty site. The potential energy of a valence bond diagram is proportional to the number of doubly occupied sites. Thus the ordinate is proportional to potential energy. The valence bond diagrams are connected via the kinetic energy operator. Paths that include valence bond diagrams with a large number of doubly occupied sites become energetically unfavourable as U is increased. Thus, quantum fluctuations between the equivalent Kekulé diagrams are reduced, thereby increasing the strength of the dimerization. Modified from figure 9 with permission from S. N. Dixit and S. Mazumdar, *Phys. Rev. B* **29**, 1824, 1984. Copyright 1984 by the American Physical Society.

$$H = \sum_n J_n \mathbf{S}_n \cdot \mathbf{S}_{n+1} + \frac{1}{4\pi t\lambda} \sum_n \Delta_n^2 + \Gamma \sum_n \Delta_n, \qquad (7.14)$$

where (to first order in Δ_n),

$$J_n = J_0 \left(1 + \frac{\Delta_n}{t} \right), \qquad (7.15)$$

and

$$J_0 = \frac{4t^2}{U - V_1}. \qquad (7.16)$$

(V_1 is the nearest neighbour Coulomb interaction). Then the bond order parameter is given by,

$$\Delta_n = -2\pi t\lambda \left(\frac{J_0}{t} \langle \mathbf{S}_n \cdot \mathbf{S}_{n+1} \rangle + \Gamma \right). \qquad (7.17)$$

Since the nearest neighbour spin-spin correlator, $\langle \mathbf{S}_n \cdot \mathbf{S}_{n+1} \rangle$, is O(1) and $J_0 \sim U^{-1}$ we see that Δ_0 decreases as a function of U in the strong-coupling limit.

This prediction is confirmed by the numerical results shown in Fig. 7.1, which agree (up to a numerical factor of ~ 2) with a strong coupling analysis in the continuum limit (Nakano and Fukuyama 1980) that predicts,

$$\Delta_0 = 8(2\pi)^{1/2}t \left(\frac{\lambda t}{U - V_1}\right)^{3/2}. \tag{7.18}$$

Also shown in Fig. 7.1 is the optical gap, defined as the *vertical* excitation energy of the lowest dipole-allowed singlet (the $1^1B_u^-$ state). This increases very rapidly with U because of a combination of two factors. First, as already discussed, the dimerization gap is increasing rapidly with U for $U < 12.5$ eV and second, the optical gap also increases with U for fixed Δ, particularly for $U > 10$ eV (as shown in Fig. 5.2). In the intermediate-coupling regime the optical gap is $30 - 40$ times larger than its noninteracting value. This behaviour illustrates a dramatic failure of the noninteracting description, as first pointed out by Ovchinnikov and coworkers (Ovchinnikov *et al.* 1973). It further illustrates the combined effects of both electron-electron and electron-phonon interactions. This will be further demonstrated in the next section when we describe the excited state structures.

7.4 Excited states and soliton structures

Electron-phonon interactions in the absence of Coulomb interactions lead to mid-gap states and associated geometric lattice defects, or solitons. In this section we explore how these geometric defects change as a function of the electron-electron interaction strength.

First, we examine the relaxed and vertical energies of the Pariser-Parr-Pople-Peierls model as a function of the interaction strength. These transition energies are illustrated in Fig. 7.4. We first note the crossover in the *vertical* energies of the $1^1B_u^-$ and $2^1A_g^+$ states as a function of U (as already discussed in Chapter 5) signifying the highly correlated nature of the $2^1A_g^+$ state at strong-coupling (Schulten and Karplus 1972). The relaxed energy of the $2^1A_g^+$ state, however, is close to or lower than that of the relaxed energy of the $1^1B_u^-$ state. For the parameter region around $0 \leqslant U \lesssim 5$ eV the *relaxed* energy of the $2^1A_g^+$ state lies slightly higher than that of the $1^1B_u^-$ state, as illustrated by the inset to Fig. 7.4. This shows the relaxed energies of the $1^1B_u^-$ and $2^1A_g^+$ states as a function of inverse chain length when $U = 2.5$ eV. [33]

For all parameter values the relaxation energy of the $1^1B_u^-$ state is modest, in contrast to the large relaxation energies of the $1^3B_u^+$ and $2^1A_g^+$ states relative to their vertical energies. These differences in the sizes of the relaxation energy is also reflected in the significant geometrical distortions from the ground state geometry in the $1^3B_u^+$ and $2^1A_g^+$ states, as described below.

---

[33]This, however, is an 'artefact' of the adiabatic approximation, which leads to self-trapping (as described in Section 7.7). This is illustrated by the inset to Fig. 7.4, which shows the relaxed energies of the $1^1B_u^-$ and $2^1A_g^+$ states as a function of inverse chain length, N^{-1}. As $N \to \infty$ the energies become constant as a consequence of self-trapping.

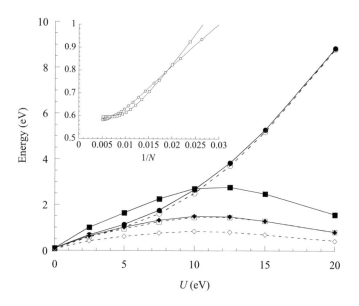

F𝗂ɢ. 7.4. The vertical (solid symbols and curves) and relaxed (open symbols and dashed curves) energies of the $1^1B_u^-$ state (circles), the $1^3B_u^+$ state (diamonds) and the $2^1A_g^+$ state (squares). The inset shows the relaxed energies of the $1^1B_u^-$ and $2^1A_g^+$ states as a function of inverse chain length when $U = 2.5$ eV. $t = 2.5$ eV and $\lambda = 0.1$.

7.4.1 $1^1B_u^-$ state

For any electronic interaction strength on a rigid, dimerized lattice this state is a bound particle-hole pair, being the $n = 1$ Mott-Wannier singlet exciton at weak-coupling and the $n = 1$ Mott-Hubbard exciton at strong-coupling. In contrast, in the noninteracting limit electron-lattice coupling results in a soliton-antisoliton pair of particle-hole spinless objects, as described by eqn (4.18) and illustrated in Fig. 4.8(a). As there is no residual attraction between the particle and hole in the noninteracting limit, they are widely separated. However, an infinitesimal particle-hole attraction binds the particle-hole pair, causing a strongly bound state as the electron interactions are increased. This behaviour is illustrated in Fig. 7.5(a), which shows the staggered bond dimerization, δ_n, (defined in eqn (4.27)) as a function of the Coulomb interaction, U. The figure clearly illustrates the 'polaronic' nature of the exciton for any nonzero Coulomb interaction, as first predicted by Grabowski *et al.* (1985). As the interaction strength increases the particle and hole become more strongly bound, the particle-hole separation decreases and the exciton creates a localized distortion of the lattice from the ground state structure. The distortion is a tendency to reduce the amplitude of the bond alternation, and is reminiscent of the polaronic distortion of a doped particle described in Section 7.5.

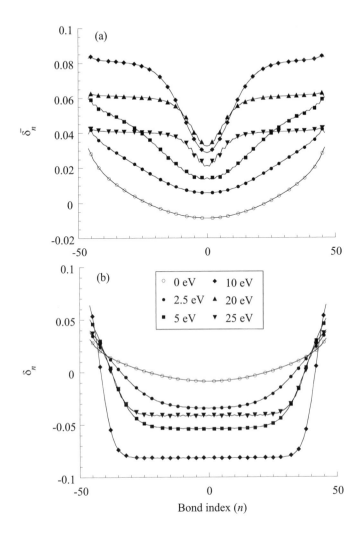

FIG. 7.5. The staggered, normalized bond dimerization, δ_n, as a function of bond index, n, of (a) the $1^1B_u^-$ state and (b) the $1^3B_u^+$ state for different values of U. The values of U are shown in the key, $t = 2.5$ eV and $\lambda = 0.1$. The calculations were performed on the Pariser-Parr-Pople-Peierls model for a 102-site chain using the DMRG method. A two-point average was performed in (a), that is, $\bar{\delta}_n = (\delta_n + \delta_{n+1})/2$.

By fitting the solitonic structures of the $1^1B_u^-$ state to the two-soliton functional form,

$$\delta_n = \delta_0 \left[1 + \tanh\left(\frac{2n_0 a}{\xi}\right) \left\{ \tanh\left(\frac{(n-n_0)a}{\xi}\right) - \tanh\left(\frac{(n+n_0)a}{\xi}\right) \right\} \right],$$

$$(7.19)$$

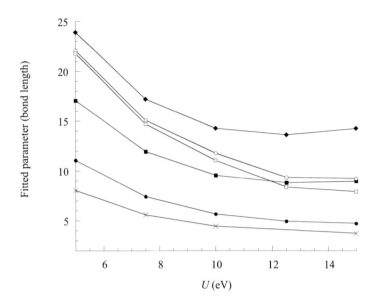

FIG. 7.6. The fitted-values of ξ (circles), $2n_0$ (squares) and $2n_d$ (diamonds) for the $1^1 B_u^-$ state (open symbols) using eqn (7.19) and the $2^1 A_g^+$ state (filled symbols) using eqn (7.20). The crosses represent the fitted values of ξ for the $1^3 B_u^+$ state. $t = 2.5$ eV and $\lambda = 0.1$.

the change in the geometrical structures as a function of U can be quantified. The correlation length, ξ, and the soliton-antisoliton separation, $2n_0$, are plotted in Fig. 7.6. This illustrates the decrease in the soliton-antisoliton separation as the interaction strength is increased.

7.4.2 $1^3 B_u^+$ state

In contrast to the $1^1 B_u^-$ state, on a rigid, dimerized lattice the $1^3 B_u^+$ state evolves from the $n = 1$ Mott-Wannier triplet exciton at weak-coupling to a pair of confined spinons at strong coupling. Thus, it acquires 'covalent' character as the electronic interactions are increased. However, as described by eqn (4.36) and illustrated in Fig. 4.8(b), in the noninteracting limit electron-lattice coupling creates a soliton-antisoliton pair of spin-1/2 objects. In this limit, therefore, the triplet state already has 'covalent' character. The effect of any electron-electron interaction is to increase this covalency, leading to a strong lattice deformation around the solitons, as shown in Fig. 7.5(b). We also see that there is a weak repulsion between the soliton-antisoliton pair, resulting in them being repelled to the ends of the chain, with a complete reversal of the bond alternation between them. The fitted correlation length, ξ, of the $1^3 B_u^+$ state, shown in Fig. 7.6, is considerably smaller than that of the $1^1 B_u^-$ state.

The $1^1 B_u^-$ and $1^3 B_u^+$ states are schematically illustrated in Fig. 7.7. We note

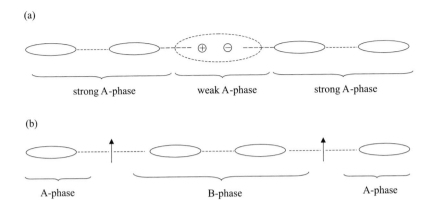

FIG. 7.7. A schematic representation of (a) the $1^1 B_u^-$ state, showing the exciton self-localized in the middle of the chain, and (b) the $1^3 B_u^+$ state, showing the spinons with a region of bond reversal between them. In the weak-coupling limit the ovals represent strong bond-order (or 'double' bonds), while the dashed lines represent weak bond-order (or 'single' bonds). In the strong-coupling limit the ovals represent singlet dimers on the 'double' bonds. The electron and hole in the singlet are strongly attracted, while the spinons in the triplet are weakly repulsed.

that the character of the $1^1 B_u^-$ and $1^3 B_u^+$ states in the weak-coupling limit are changed by the electron-lattice coupling. On a rigid geometry the vertical excitations are described as particle-hole excitations from the delocalized HOMO to the delocalized LUMO. The particle and hole form a bound state (or exciton) as a consequence of the attractive particle-hole potential. However, the relaxed $1^1 B_u^-$ state is more conveniently described as a pair of spinless oppositely charged particles occupying the localized Wannier orbitals associated with the mid-gap states ($S^\pm$). Electron-hole interactions also bind the oppositely charged solitons to form an exciton. In contrast, the $1^3 B_u^+$ state is more conveniently described as a pair of neutral spinons occupying these localized Wannier orbitals (S^σ). Being neutral, the spinons do not form a bound state. It is also clear - both from the geometrical structures and the relaxation energies - that the coupled effect of electron-electron and electron-lattice interactions is more significant for the $1^3 B_u^+$ state than for the $1^1 B_u^-$ state. The reason for this is that the bond-order operator (defined in eqn (7.2)) couples to the covalent character of a state, which is greater for the $1^3 B_u^+$ state than for the $1^1 B_u^-$ state.

7.4.3 $2^1 A_g^+$ state

Electron-lattice coupling also has a rather dramatic affect on the $2^1 A_g^+$ state, again because of its 'covalent' character. On a rigid, dimerized lattice in the weak-coupling limit this state is the $n = 2$ singlet Mott-Wannier exciton, evolving to a pair of spin-1 objects at strong-coupling. Since the spin-1 objects are themselves

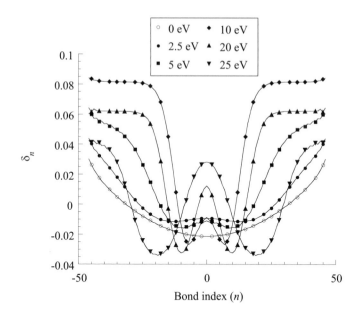

FIG. 7.8. The staggered, normalized bond dimerization, δ_n, as a function of bond index, n, of the $2^1 A_g^+$ state for different values of U. The values of U are shown in the key, $t = 2.5$ eV and $\lambda = 0.1$. Singlet fission into two triplets is illustrated by the geometric structure at $U = 25$ eV.

comprised of a pair of spin-1/2 objects (as illustrated in Fig. 5.3), it should not be surprising to find that the $2^1 A_g^+$ state is described by four-solitons in the strong-coupling limit (Hayden and Mele 1986; Su 1995). Figure 7.8 indeed shows that a four-soliton description is relevant for all nonzero interaction strengths, becoming more pronounced as the Coulomb interaction increases.

The four-soliton form (Su 1995),

$$\delta_n = \delta_0 [1 + \tanh\left(\frac{2n_0 a}{\xi}\right) \{\tanh\left(\frac{(n - n_d - n_0)a}{\xi}\right) - \tanh\left(\frac{(n - n_d + n_0)a}{\xi}\right)$$
$$+ \tanh\left(\frac{(n + n_d - n_0)a}{\xi}\right) - \tanh\left(\frac{(n + n_d + n_0)a}{\xi}\right)\}], \tag{7.20}$$

can be used to extract the soliton parameters of the $2^1 A_g^+$ state. We interpret $2n_0$ as the distance between the soliton-antisoliton pair and $2n_d$ is the distance between the centre-of-masses of the pair of solitons pairs. Their U dependence is shown in Fig. 7.6, showing enhanced soliton confinement as U is increased up to the intermediate regime.

For interactions less than a critical value of U the soliton parameters are independent of chain length, indicating a confinement of the solitons, both within a pair and between pairs. However, at a critical value of U the four solitons

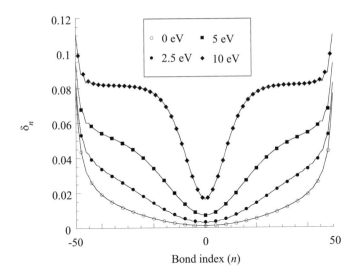

FIG. 7.9. The staggered, normalized bond dimerization, δ_n, as a function of bond index, n, of the doped (polaron) state for different values of U. The values of U are shown in the key, $t = 2.5$ eV and $\lambda = 0.1$.

become unconfined with $2n_0 \sim N/4$ and $2n_d \sim N/2$. Qualitatively, we can understand this by recalling from Chapter 4 that at strong-coupling the $2^1 A_g^+$ state may be viewed as a pair of spin-1 objects, which are unbound except at a vanishingly small dimerization. This singlet fission into a pair of triplets occurs at large values of U ($U > 22.5$ eV), and is illustrated in Fig. 7.8 by the geometrical structures for $U = 25$ eV.

7.5   Polarons

A polaron is charged particle associated with a lattice distortion. These were described in the noninteracting limit in Chapter 4, where their origins arise solely via the coupling of the lattice to the bond-order operator. In the interacting limit there is an additional coupling to the lattice via the Coulomb interaction. Figure 7.9 shows the polaronic distortions for various interaction strengths. The lattice distortions for nonzero interactions are qualitatively similar to those of the $1^1 B_u^-$ state, illustrated in Fig. 7.5, confirming the exciton-polaron character of that state. However, as described later in Section 7.7, because the charged particle has both short-range interactions with acoustic phonons and long-range interactions with longitudinal optic phonons, the coupling to the lattice is somewhat larger than that for the exciton-polaron. This is demonstrated by the narrower and deeper distortion for the polaron relative to the exciton-polaron.

7.6   Extrinsic dimerization

We have seen in this chapter the dramatic differences the combined effects of electronic interactions and electron-lattice coupling have on the character and geometrical structures of the lowest singlet and triplet excited states: electron-lattice coupling enhances the 'ionic' character of the $1^1B_u^-$ state, whereas it enhances the 'covalent' character of the $1^3B_u^+$ state. These observations apply to linear chains in the absence of extrinsic dimerization. The question therefore arises as to how the character of excited states of an extrinsically semiconducting polymer, as for light emitting polymers or polydiacetylene, for example, evolve as a function of Coulomb interactions for a fixed electron-lattice coupling.

We can begin to address this question by modelling an extrinsically semiconducting polymer as a linear chain with an extrinsic dimerization. This is achieved by the inclusion of t_e in eqn (7.3), as described in Section 4.8. As in the absence of this term, in the noninteracting limit electron-lattice coupling causes mid-gap electronic states, and associated localized soliton wavefunctions and geometrical defects. The new affect of the extrinsic dimerization is to generate a linear confining potential between the soliton-antisoliton pairs. This new property of soliton-antisoliton confinenment is illustrated by the localized Wannier orbitals associated with the soliton, ϕ_n, and antisoliton, $\bar{\phi}_n$. These are obtained from the molecular orbitals associated with the mig-gap electronic states, $\psi_n^{\pm}$.

Figure 4.11 shows the probability density of the Wannier orbitals associated with the mid-gap states. Although the relative separation of Wannier orbitals is small with an extrinsic dimerization of $\delta_e = 0.1$, the fact that there are two distinct Wannier orbitals implies that the argument employed in Section 4.6 - concerning the different characters of the $1^1B_u^-$ and $1^3B_u^+$ states after electron-lattice relaxation - is a general one. Thus, the $1^1B_u^-$ state is comprised of spinless electron-hole pairs, while the $1^3B_u^+$ state is comprised of two spin-1/2 objects. These become confined in the presence of extrinsic dimerization. We would therefore expect that, as before, the different character of the $1^1B_u^-$ and $1^3B_u^+$ states will be evident by the different type of geometrical distortions when electron-electron interactions are included.

Figure 7.10(a) and (b) shows the geometric structures of the $1^1B_u^-$ and $1^3B_u^+$ states, respectively. As the extrinsic dimerization causes a confinement of the soliton-antisoliton pair, the geometrical structures are 'polaronic' in the noninteracting limit for both cases. For the $1^1B_u^-$ state, as before, increased Coulomb interactions bind the particle-hole pair into an exciton, resulting in very little change to the geometrical structure. For the $1^3B_u^+$ state, however, electronic interactions lead to a more pronounced change in the geometrical structure, resulting in a reversal in the intrinsic bond dimerization for a moderately small U. We therefore see that the qualitative picture of the $1^1B_u^-$ and $1^3B_u^+$ states illustrated in Fig. 7.7 in the absence of extrinsic dimerization remains essentially valid here, except that now the two spinons are bound in the $1^3B_u^+$ state and there is a weaker reversal in the intrinsic bond alternation between them.

The geometrical structure of the $2^1A_g^+$ state is shown in Fig. 7.11. Notice that

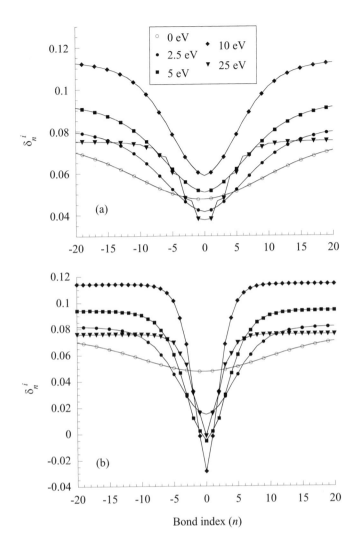

FIG. 7.10. The staggered, normalized, intrinsic bond dimerization, δ_n^i, as a function of bond index, n, for a linear chain with extrinsic dimerization, $\delta_e = 0.1$. (a) $1^1B_u^-$ state and (b) $1^3B_u^+$ state. Notice that the total bond dimerization, $\delta_n = \delta_n^i + \delta_e$, does not change sign for the triplet state. $t = 2.5$ eV. $\lambda = 0.1$.

here, as for the $1^3B_u^+$ state, there is a change of sign of the intrinsic dimerization in the middle of the chain. However, now the $2^1A_g^+$ state fits a two-soliton form for $U \gtrsim 2.5$ eV, unlike the four-soliton form in the absence of extrinsic dimerization.

The greater lattice distortions of the $1^3B_u^+$ and $2^1A_g^+$ states relative to the $1^1B_u^-$ state is also reflected in their larger energy relaxations. These are listed in Table 7.1 for a 66-site chain. Notice that there is a vertical gap of ca. 1 eV

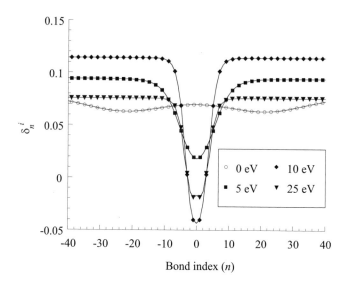

FIG. 7.11. The staggered, normalized, intrinsic bond dimerization, δ_n^i, as a function of bond index, n, of the $2^1A_g^+$ state for a linear chain with extrinsic dimerization, $\delta_e = 0.1$. $t = 2.5$ eV, and $\lambda = 0.1$.

Table 7.1 *The vertical and relaxation energies (in eV) of a 66 site linear polyene calculated from the Pariser-Parr-Pople-Peierls model with extrinsic dimerization ($t = 2.5$ eV, $U = 10.0$ eV, $\lambda = 0.1$, and $\delta_e = 0.1$)*

State	Vertical transition energy	Relaxation energy
$1^3B_u^+$	2.95	0.37
$1^1B_u^-$	4.12	0.14
$2^1A_g^+$	5.07	0.91

between the $1^1B_u^-$ and $2^1A_g^+$ states, but their relaxed energies are rather close.

7.7 Self-trapping

Within the Born-Oppenheimer appproximation we have seen that the excited and charged states become self-localized, or self-trapped (Holstein 1959). Self-trapping in one-dimensional systems is a consequence of the well-known result from quantum mechanics that a symmetric attractive potential has at least one bound state (see Song and Williams (1993) for more details and references). In this case the attractive potential is the relaxation energy of the excited or charged state associated with the deformation of the lattice.

Within a semiclassical, continuum approximation (Emin and Holstein 1976; Toyozawa and Shinozuka 1980) the energy of the self-trapped state as a function of the variable, $\gamma = r_0/r$, where r is the spatial extent of the centre-of-mass of the electronic wavefunction and r_0 is the lattice parameter, is

$$E(\gamma) = 2t(\gamma^2 - g_s\gamma - g_\ell\gamma). \tag{7.21}$$

g_s, representing the short-range interaction with the acoustic phonons, is

$$g_s = \frac{E_d^2}{4tKr_0^3}, \tag{7.22}$$

where E_d is the relaxation energy. g_ℓ, representing the long-range interaction with the longitudinal optic phonons, is

$$g_\ell = \frac{e^2}{2t\tilde{\epsilon}r_0}, \tag{7.23}$$

where $1/\tilde{\epsilon} = 1/\epsilon_\infty - 1/\epsilon_0$, and ϵ_∞ and ϵ_0 are the high frequency and static dielectric constants, respectively. Thus, the equilibrium spatial extent of the electronic state, r_{eq}, is

$$r_{eq} = \frac{2r_0}{g_s + g_\ell}. \tag{7.24}$$

Since the excited states are electrically neutral, only the short-range, acoustic interaction is relevant in eqn (7.24). (This is also true for the exciton-polaron, as the particle and hole are closely separated.) The polaron, however, being charged also couples to the longitudinal optic phonons, so the long-range term is retained in eqn (7.24).

As we have already discussed in Section 4.11, self-trapping can never be a true consequence of electron-phonon interactions in a translationally invariant Hamiltonian: it is an artefact of the adiabatic approximation, which freezes the nuclear degrees of freedom. When the nuclear degrees of freedom are quantized it is possible to construct a translationally invariant wave-packet of both the electron and nuclear degrees of freedom. The band width of this wavepacket is a function of the phonon frequency, and in the adiabatic limit ($\omega \to 0$) it will vanish.[34]

Of course, defects or other imperfections destroy translational invariance, and these are always present in one-dimensional systems. It is thus useful to describe a particle as practically self-trapped if the experimental timescales are shorter than the time taken for the particle to tunnel out of its localized state.

7.8 Concluding remarks

We conclude this chapter by summarizing the combined effects of electron-electron and electron-lattice interactions on the electronic states of conjugated polymers.

---

[34]Quantum corrections to the adiabatic limit will be discussed in Section 10.3.

- Generally, electronic interactions considerably enhance the bond alternation, particularly at the physically relevant values of $U \sim 4t$. However, this enhancement diminishes in the strong coupling limit.

- The enhancement of the bond alternation, coupled to the effect of electron interactions on the optical gap, means that the optical gap is considerably enhanced from its noninteracting value.

- The $1^1B_u^-$ state is comprised of a pair of spinless oppositely charged particles, and forms an exciton-polaron.

- The $1^3B_u^+$ state is comprised of two spin-1/2 spinons, which weakly repel in the absence of extrinsic dimerization.

- The $2^1A_g^+$ state has a four-soliton character. In the strong-coupling limit there is singlet fission to a pair of triplets.

- There is a substantial energy relaxation for the $1^3B_u^+$ and $2^1A_g^+$ states, but rather modest energy relaxation for the $1^1B_u^-$ state. This is a consequence of the more covalent character of the $1^3B_u^+$ and $2^1A_g^+$ states in comparison to the $1^1B_u^-$ state, and because the bond-order operator couples to the covalent character of a state. Thus there can be energy level reversal, with the relaxed energy of the $2^1A_g^+$ state being near to or below that of the $1^1B_u^-$ state.

- These features remain qualitatively correct when there is extrinsic dimerization, except that the soliton-antisoliton pairs are confined for both the $1^1B_u^-$ and $1^3B_u^+$ states, and the $2^1A_g^+$ state has a two-soliton character.

The predictions presented in this chapter are all within the Born-Oppenheimer approximation. Quantum phonons will reduce the amplitude of the bond alternation in the ground state (Fradkin and Hirsch 1983; McKenzie and Wilkin 1992) - for realistic models of *trans*-polyactylene by about 20% (Barford *et al.* 2002b), as described in Chapter 10. Quantum phonons also prevent self-trapping. This latter has a rather significant affect on the relaxed energies of the $1^3B_u^+$ and $2^1A_g^+$ states in linear polyenes, as explained in Chapter 10.

In this chapter we have drawn together the effects of electron-electron and electron-phonon coupling as a function of parameter space. In Chapter 10 we consider the specific physical example of *trans*-polyacetylene, while in Chapter 11 we focus on the phenyl-based light emitting polymers. The next two chapters describe how electron-lattice relaxation plays a key role in determining the optical and electronic processes in conjugated polymers.

8

OPTICAL PROCESSES IN CONJUGATED POLYMERS

8.1 Introduction

There are many excellent accounts of both the theory of optical processes in general (Ziman 1972; Butcher and Cotter 1990; Mukamel 1995 and Loudon 2000), and optical processes in organic (Pope and Swenberg 1999) and inorganic materials (Henderson and Imbusch 1989), in particular. It is the purpose of this chapter to describe some of the important linear and nonlinear optical processes that enable us to establish a connection between the theories of electronic states described in this book and their experimental consequences.

Much of the recent interest in conjugated polymers has been inspired by the optical properties of the light emitting phenyl-based systems. Unlike *trans*-polyacetylene, the phenyl-based systems luminesce, because the lowest-lying singlet excited state is dipole connected to the ground state. As we describe in Chapters 10 and 11, this is a consequence of the different electron-electron and electron-phonon interactions in these two types of system. Another potentially important application of conjugated polymers is in nonlinear optical devices, which exploit the fact that the polarizability depends nonlinearly on the electric field. Such devices include optical switches, frequency multipliers and electric-optic modulators (Pope and Swenberg 1999).

Linear optical processes give important information about the energies of the dipole allowed states. However, 'dark' states - namely those with no dipole moment to the ground state - are inaccessible. Nonlinear optical processes, on the other hand, involve transitions between two or more states, so these access the dipole-forbidden states. In this chapter we explain how third order nonlinear process can be used to identify these forbidden states.

Consider a system of N polymers per unit volume under the influence of driving electric fields $\mathbf{E}(\omega_1)$, $\mathbf{E}(\omega_2)$, $\mathbf{E}(\omega_3)$, etc. Then the response of a system at a frequency $\omega_\sigma = \omega_1 + \omega_2 + \omega_3 + \cdots$, as measured by its polarization, $\mathbf{P}(\omega_\sigma)$, is,

$$\mathbf{P}(\omega_\sigma) = \chi^{(1)}(-\omega_\sigma; \omega_1)\mathbf{E}(\omega_1) + \chi^{(2)}(-\omega_\sigma; \omega_1, \omega_2)\mathbf{E}(\omega_1)\mathbf{E}(\omega_2)$$
$$+ \chi^{(3)}(-\omega_\sigma; \omega_1, \omega_2, \omega_3)\mathbf{E}(\omega_1)\mathbf{E}(\omega_2)\mathbf{E}(\omega_3) + \cdots. \qquad (8.1)$$

The nth order electrical susceptibility, $\chi^{(n)}$, is an $n+1$ rank tensor. If we rewrite eqn (8.1) as,

$$\mathbf{P} = \chi_{\text{eff}}^{(1)}(\mathbf{E})\mathbf{E}, \qquad (8.2)$$

we see that there is an effective electric field-dependent linear susceptibility, and thus electric field-dependent refractive indices, for example.

Before discussing the nonlinear effects we first establish our notation by discussing linear optical properties.

8.2   Linear optical processes

For transitions from the ground state, $|0\rangle$, to the excited states $\{|J\rangle\}$ with energies $\{E_J\}$ the first order susceptibility is

$$\chi_{\alpha\beta}^{(1)}(\omega) = \frac{N}{\epsilon_0 \hbar} \sum_J \left[\frac{\langle 0|\hat{\mu}_\alpha|J\rangle\langle J|\hat{\mu}_\beta|0\rangle}{\Omega_J - \omega} + \frac{\langle 0|\hat{\mu}_\beta|J\rangle\langle J|\hat{\mu}_\alpha|0\rangle}{\Omega_J + \omega} \right]. \tag{8.3}$$

$\Omega_J = (E_J - E_0)/\hbar$ is the angular transition frequency of the state $|J\rangle$ and $\hat{\mu}_\alpha$ is the αth cartesian component of the dipole operator, $\hat{\mu}$.

If the polymers are oriented along the x-axis, the dominant susceptibility is $\chi_{xx}^{(1)}$. Then, denoting $\chi_{xx}^{(1)}$ as $\chi^{(1)}$ we have

$$\chi^{(1)}(\omega) = \frac{N}{\epsilon_0 \hbar} \sum_J \left[\frac{\langle 0|\hat{\mu}_x|J\rangle\langle J|\hat{\mu}_x|0\rangle}{\Omega_J - \omega} + \frac{\langle 0|\hat{\mu}_x|J\rangle\langle J|\hat{\mu}_x|0\rangle}{\Omega_J + \omega} \right], \tag{8.4}$$

$$= \frac{Ne^2}{m\epsilon_0} \sum_J \frac{f_J}{\Omega_J^2 - \omega^2}, \tag{8.5}$$

where

$$f_J = \frac{2m}{e^2 \hbar} \Omega_J \langle 0|\hat{\mu}_x|J\rangle^2 \tag{8.6}$$

is the oscillator strength for the transition from $|0\rangle$ to $|J\rangle$. The oscillator strength satisfies the important sum rule that

$$\sum_J f_J = N_e, \tag{8.7}$$

where N_e is the number of π-electrons in the polymer.

The linear optical properties follow directly from $\chi^{(1)}$. For example, the bulk dielectric function (or relative permittivity), $\epsilon(\omega)$, is

$$\epsilon(\omega) = 1 + \chi^{(1)}(\omega). \tag{8.8}$$

Then, the linear absorption coefficient, defined as the fraction of energy absorbed in passing through a unit thickness of material, is

$$\alpha(\omega) = \frac{\omega}{nc} \text{Im}[\epsilon(\omega)]$$

$$= \frac{\omega}{nc} \text{Im}[\chi^{(1)}(\omega)]$$

$$= \frac{\pi\omega N}{nc\epsilon_0 \hbar} \sum_J \langle 0|\hat{\mu}_x|J\rangle^2 \delta(\omega - \Omega_J), \tag{8.9}$$

where n is the refractive index and c is the speed of light.[35]

---

[35]ω is implicitly complex, as it contains an imaginary term to represent damping.

The time scale for an optical transition to the state $|J\rangle$ is $\sim 2\pi/\Omega_J \sim 10^{-15}$ s for transitions in the visible region. Once excited to the state $|J\rangle$ there is rapid nonradiative interconversion to the lowest excited singlet state. The interconversion may be a multiphonon process - arising from the coupling of the Born-Oppenheimer states via the nonadiabatic Hamiltonian (eqn (2.19)), or single-phonon processes - arising from the coupling of the system to a phonon bath (DiBartolo 1980). If the lowest excited singlet state is dipole connected to the ground state, then in general there is both radiative spontaneous emission and nonradiative emission to the ground state. The radiative inverse life-time, Γ_J, is given by the Einstein expression

$$\Gamma_J \equiv \tau^{-1} = \frac{n(\hbar\Omega_J)^3 \langle 0|\hat{\mu}_x|J\rangle^2}{\pi\epsilon_0\hbar^4c^3}. \tag{8.10}$$

Typically, the life-time of the lowest excited state in polymers is $\sim 10^{-10} - 10^{-9}$ s.

This behaviour after photoexcitation is encapsulated by the following empirical rules (Birks 1970).

- *Vavilov's Rule*: The fluorescence quantum efficiency is independent of the excitation wavelength. This implies that there is efficient nonradiative interconversion.

- *Kasha's Rule*: Emission occurs from the lowest excited singlet state.

However, in some polymers the lowest excited singlet state is not dipole connected to the ground state, and in those cases there are only nonradiative transitions to the ground state.

8.3   Evaluation of the transition dipole moments

Evidently, the evaluation of optically important parameters, such as $\chi^{(1)}$ and Γ_J, depends on the evaluation of the transition dipole moments, $\langle 0|\hat{\mu}|J\rangle$. In calculating the dipole moments a considerable simplification arises if we adopt the Franck-Condon principle. This is discussed in the next section.

8.3.1   *The Franck-Condon principle*

A general state, $|J\rangle$, of the polymer is a function of many degrees of freedom, corresponding to the electron and nuclear coordinates. As usual, it is convenient to represent the nuclear degrees of freedom as normal modes, with each normal mode being associated with a normal coordinate, Q_α, and a characteristic frequency, ω_α. To simplify the discussion of the Franck-Condon principle we will make the reasonable assumption that only one normal mode is strongly coupled to the electronic degrees of freedom.

The Franck-Condon principle is essentially a restatement of the Born-Oppenheimer approximation (introduced in Chapter 2), as it assumes that the electronic transition occurs so quickly that the nuclear coordinates remain stationary. Phonon

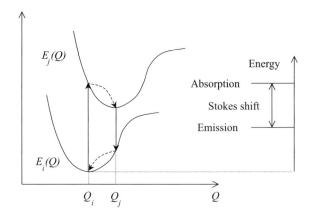

FIG. 8.1. The adiabatic energy curves of the electronic states $|i; Q\rangle$ and $|j; Q\rangle$, as a function of the normal coordinate, Q. The solid up and down vertical arrows are the vertical absorption and emission transitions, respectively. The dashed arrows represent the nonradiative vibrational relaxation. The Stokes shift is twice the reorganization energy.

frequencies are typically one order of magnitude smaller than optical transition energies, corresponding to nuclear motion times being roughly ten times longer than electronic transition times. Thus, the Born-Oppenheimer approximation is generally valid.

The Franck-Condon principle is illustrated in Fig. 8.1. The solid curves represent the adiabatic energy of the electronic states as a function of a normal coordinate, Q. Generally an excited state will have an electronic energy minimum at a different Q value than the ground state, as the electronic distributions differ in both states. Classically, the transitions are *vertical*, that is the transition occurs to the energy of the excited state with the same Q value as the ground state. This is illustrated by the up-vertical arrow. Since the life-time of the excited state is much longer than nuclear motion times, after the vertical transition there is a *relaxation* of the nuclear coordinates to the bottom of the adiabatic energy curve. The transition to the ground state is again vertical, with an emission energy less than the absorption energy. This energy difference is the *Stokes shift*.

For small displacements of Q from equilibrium the adiabatic energy profiles are quadratic, and thus fluctuations in Q may be quantized as linear harmonic oscillators. The energy of the oscillators is represented by the horizontal lines in Fig. 8.2. Thus, quantum mechanically, there is a progression of linear harmonic oscillator states for each electronic state. Vibronic transitions can occur between pairs of vibronic states.

To calculate the amplitude of these transitions we adopt the Born-Oppenheimer approximation and factorize $|J\rangle$ as a single, direct product of the electronic and nuclear degrees of freedom

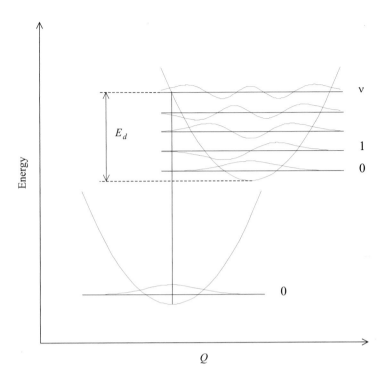

FIG. 8.2. The vibrational energy levels and the associated (unnormalized) linear harmonic oscillator wavefunctions. The Huang-Rhys parameter, $S = E_d/\hbar\omega$, and thus $S = \nu + \frac{1}{2}$. E_d is the reorganization (or relaxation) energy.

$$|J\rangle = |j; Q\rangle|\nu_j\rangle, \tag{8.11}$$

where $|j; Q\rangle$ represents the electronic part, parametrized by the normal coordinate, Q, and $|\nu_j\rangle$ represents the nuclear part associated with that electronic state.[36]

The total dipole operator is the sum of the electronic and nuclear dipole moments,

$$\hat{\mu} = \hat{\mu}_e + \hat{\mu}_N. \tag{8.12}$$

Then, the total transition dipole moment between the states $|I\rangle$ and $|J\rangle$ is,

$$\begin{aligned}
\langle I|\hat{\mu}|J\rangle &= \langle i; Q|\hat{\mu}_e|j; Q\rangle\langle\mu_i|\nu_j\rangle + \langle i; Q|j; Q\rangle\langle\mu_i|\hat{\mu}_N|\nu_j\rangle \\
&= \mu_{ij}\langle\mu_i|\nu_j\rangle,
\end{aligned} \tag{8.13}$$

because $\langle i; Q|j; Q\rangle = \delta_{ij}$, as the states $\{|i; Q\rangle\}$ are ortho-normal.

[36]We use upper case Latin letters to represent a general electron-phonon state, lower case Latin letters to represent the electronic state, and lower case Greek letters to represent the phonon states.

$$\mu_{ij} = \langle i; Q | \hat{\mu}_e | j; Q \rangle \tag{8.14}$$

is the electronic dipole moment and $\langle \mu_i | \nu_j \rangle$ is the instantaneous overlap of the nuclear wavefunctions. The matrix element in eqn (8.14) is evaluated at the same value of Q in both the initial and final electronic states, namely at the equilibrium value of Q in the initial state.

Symmetry rules dictate that the electronic matrix elements are only nonzero for electronic states of definite symmetries. We introduce these rules in the next section, followed by a discussion of the the Franck-Condon factors, which determine the intensity of the vibrational transitions. Finally, we use the exciton model (described in Chapter 6) to evaluate the electronic dipole moments.

8.3.2   Electronic selection rules

There are three important selection rules for electronic transitions owing to the properties of the electronic dipole operator, $\hat{\mu}_e$. The electronic dipole operator is defined as

$$\hat{\mu}_e = e\hat{r} = e \sum_i \mathbf{r}_i (\hat{N}_i - 1), \tag{8.15}$$

where $\mathbf{r}_i$ is the position of the ith site, $\hat{N}_i$ is the number operator, and the sum is over all atomic sites.

- The electric dipole operator conserves total spin, so transitions only occur between states in the same spin manifold.
- The electric dipole operator is antisymmetric with respect to the inversion operator, $\hat{i}$, and thus it connects states of opposite inversion symmetry. To see this note that,

$$\langle i | \hat{\mu}_e | j \rangle \equiv \langle i | \hat{i}^\dagger \hat{i} \hat{\mu}_e \hat{i}^\dagger \hat{i} | j \rangle = -\langle i | \hat{i}^\dagger \hat{\mu}_e \hat{i} | j \rangle = -i_i i_j \langle i | \hat{\mu}_e | j \rangle, \tag{8.16}$$

  where $\hat{i} | j \rangle = i_j | j \rangle$, i_j is the eigenvalue of $\hat{i}$ and $\hat{i} \hat{\mu}_e \hat{i}^\dagger = -\hat{\mu}_e$. Thus, $\langle i | \hat{\mu}_e | j \rangle$ is nonzero only if $i_i i_j = -1$.
- The electric dipole operator is antisymmetric with respect to the particle-hole operator and thus it connects states of opposite particle-hole symmetry. The proof is identical to that for the inversion operator.

Centro-symmetric polymers, whose Hamiltonians are invariant under $\hat{i}$, have states classified as A_g (even) and B_u (odd). The ground state is a singlet A_g state, so transitions between the ground state and singlet B_u states occur, but not between the ground state and other A_g states. Transitions between the ground state and triplet states are forbidden. (However, if there is spin-orbit coupling, the Hamiltonian eigenstates are not eigenstates of total spin, the 'triplet' states will contain some singlet character, and there will be phosphorescence from the lowest triplet state.)

8.3.3 Franck-Condon factors

The Franck-Condon factor, $F_{\mu\nu}$, is the square magnitude of the instantaneous overlap of the nuclear wavefunctions,

$$F_{\mu\nu} = \langle\mu_i|\nu_j\rangle^2. \tag{8.17}$$

As the intensity of the transitions are proportional to the square of the dipole moments (see eqn (8.6)), the Franck-Condon factors weight each of the vibronic transitions. To evaluate these terms we write

$$\langle\mu_i|\nu_j\rangle = \langle\mu_i|\left\{\int dQ|Q\rangle\langle Q|\right\}|\nu_j\rangle = \int \phi_\mu(Q-Q_i)\phi_\nu(Q-Q_j)dQ, \tag{8.18}$$

where

$$\phi_\mu(Q-Q_i) \equiv \langle Q|\mu_i\rangle \tag{8.19}$$

is the μth linear harmonic wavefunction centred at Q_i.

The overlap integral may be expressed as (Keil 1965),

$$\langle\mu_i|\nu_j\rangle = \sqrt{\frac{\mu!}{\nu!}}\left(-\frac{A}{\sqrt{2}}\right)^{(\nu-\mu)}\exp(-A^2/4)L_\mu^{\nu-\mu}(A^2/2), \tag{8.20}$$

where $L_m^n(x)$ are the associated Laguerre polynomials,

$$L_m^n(x) = \sum_{k=0}^{m}\frac{(-1)^k(m+n)!}{(m-k)!(n+k)!k!}x^k, \tag{8.21}$$

and

$$A = \sqrt{\frac{M\omega}{\hbar}}(Q_i - Q_j), \tag{8.22}$$

is the difference in the dimensionless electron-lattice coupling between the electronic states $|i\rangle$ and $|j\rangle$.

At $T = 0$ K only the lowest vibrational state ($\mu = 0$) of the ground state ($|I\rangle$) is occupied, and we define the zero-temperature Franck-Condon factor as

$$\begin{aligned} F_{0\nu} = \langle 0|\nu\rangle^2 &= \frac{\exp(-A^2/2)(A^2/2)^\nu}{\nu!} \\ &= \frac{\exp(-S)S^\nu}{\nu!}, \end{aligned} \tag{8.23}$$

where S is the Huang-Rhys parameter (Huang and Rhys 1950), defined by

$$S = \frac{A^2}{2} = \frac{M\omega}{2\hbar}(Q_i - Q_j)^2. \tag{8.24}$$

The Huang-Rhys parameter has a useful, physical interpretation:

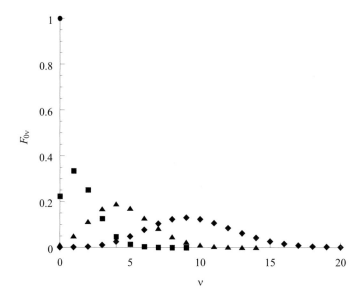

FIG. 8.3. The zero-temperature Franck-Condon factor, $F_{0\nu}$, as a function of ν for different values of the Huang-Rhys parameter, S. $S = 0$ (circles), $S = 1.5$ (squares), $S = 4.5$ (triangles), and $S = 9.5$ (diamonds). The overlap of the harmonic oscillator wavefunctions, shown in Fig. 8.2, ensures that the $0 - \nu$ (or vertical) transition is the largest.

$$S\hbar\omega = E_d = \left(\nu + \frac{1}{2}\right)\hbar\omega, \tag{8.25}$$

where E_d and ν are defined in Fig. 8.2. Thus,

$$S = \nu + \frac{1}{2}, \tag{8.26}$$

where ν is the nearest vibrational level to which a vertical transition from the $\mu = 0$ ground state level reaches. E_d is the reorganization (or relaxation) energy. Figure 8.3 shows $F_{0\nu}$ for different values of S. We note that

- Only 0-0 transitions occur when $S = 0$.
- $F_{0\nu}$ satisfies the sum rule, $\sum_\nu F_{0\nu} = 1$. So, oscillator strength is transferred from the 0-0 transition to higher transitions as S increases.
- The dominant transition is to the $|\nu\rangle$ vibrational state, where $\nu = S - 1/2$. We therefore see that the vertical, classical transition dominates.
- In general, $F_{0\nu}$ is a Poisson distribution. However, as S increases the profile of $F_{0\nu}$ becomes a Gaussian function of ν.
- S may be obtained empirically from the experimental vibronic progression by noting from eqn (8.23) that $S = F_{01}/F_{00}$.

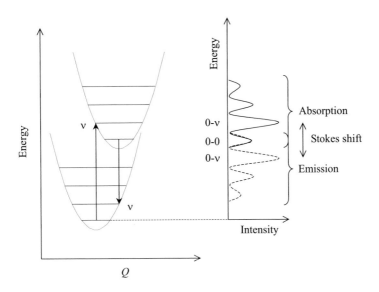

FIG. 8.4. The adiabatic energy curves for the initial and final states, the vibrational energy levels, the vertical absorption and emission transitions, and the associated intensity of the absorption and emission spectra determined by the Franck-Condon factors.

Figure 8.4 summarizes Figs 8.2 and 8.3 by showing the vibronic transitions and the optical spectra associated with these transitions determined by the Franck-Condon factors. A well-defined vibronic progression indicates that the excited and ground states have a nonzero Huang-Rhys parameter, and thus they have different geometries. This usually implies, but does not prove, that the excited state is self-trapped.

8.3.4 Electronic dipole moments: Application of the exciton model

Having discussed the vibrational overlaps, the final task is to evaluate the electronic transition dipole moments. We obtain insight into the behaviour of the dipole moments by using the effective-particle exciton model, introduced in Chapter 6.

In the exciton model the states are expressed as,

$$|^1B_u^-\rangle \approx \sum_{\text{odd } n}\sum_{\text{odd } j} \alpha_{nj}|\Phi_{nj}\rangle \qquad (8.27)$$

and

$$|^1A_g^+\rangle \approx \sum_{\text{even } n}\sum_{\text{odd } j} \alpha_{nj}|\Phi_{nj}\rangle, \qquad (8.28)$$

where $|\Phi_{nj}\rangle$ represents a Mott-Wannier exciton eigenstate in the weak-coupling limit, or a Mott-Hubbard exciton eigenstate in the strong-coupling limit. n and

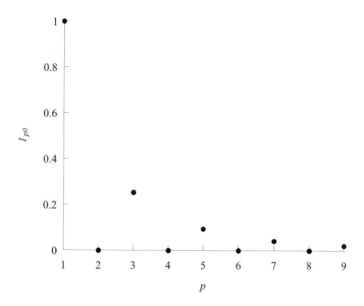

FIG. 8.5. The relative intensities, I_{0p}, (proportional to the square of the transition dipole moments) with respect to the $|0\rangle \rightarrow |1\rangle$ transition, calculated using the weak-coupling exciton theory on a linear chain. $U = 10$ eV, $t = 2.5$ eV, and $\delta = 0.2$. Transitions from the ground state to even p states are forbidden by particle-hole symmetry. $p = n_p$ and $j_p = 1$ for all cases.

j are the principle and pseudomomentum quantum numbers, respectively. In practice, only one component dominates the sums in eqns (8.27) and (8.28).

The transition dipole moments may be evaluated using the explicit expressions for $|\Phi_{nj}\rangle$, as described in Appendix F. Below we summarize the results of that calculation. In all cases the results are consistent with the selection rules described in Section 8.3.2, namely that only $A_g^+ \leftrightarrow B_u^-$ transitions are allowed.

8.3.4.1　The weak-coupling (Mott-Wannier) limit

- For transitions between the ground state and the excited state, $|p\rangle$ (with quantum numbers n_p and j_p), the transition dipole moment for a polymer oriented along the x-axis is,

$$e\langle 1^1 A_g^+ | \hat{x} | p\rangle \sim e\sqrt{\frac{L}{r_p} \frac{d}{j_p}}, \tag{8.29}$$

for odd n_p and odd j_p, and

$$e\langle 1^1 A_g^+ | \hat{x} | p\rangle = 0, \tag{8.30}$$

otherwise.

r_p is the root-mean-square particle-hole separation of the state $|p\rangle$, L is the length of the chain and d is the unit cell repeat distance.

We see that $\langle 1^1 A_g^+ | \hat{x} | p \rangle \propto \sqrt{L/r_p}$, and thus the oscillator strength is largest for the most strongly bound exciton. The dependence on L is a consequence of the sum rule, eqn (8.7).

Figure 8.5 shows the relative intensities (proportional to the square of the transition dipole moments) for transitions from the ground state to a number of excited states, calculated using the exciton theory.

- For transitions between two excited states, $|p\rangle$ and $|q\rangle$, the transition dipole moment is

$$e\langle p | \hat{x} | q \rangle = e \int dr \psi_{n_p}(r) r \psi_{n_q}(r), \qquad (8.31)$$

for $|n_p - n_q| = $ odd and $j_p = j_q$, and

$$e\langle p | \hat{x} | q \rangle = 0, \qquad (8.32)$$

otherwise.

$\psi_{n_p}(r)$ and $\psi_{n_q}(r)$ are the effective-particle wavefunctions for the states $|p\rangle$ and $|q\rangle$, respectively, and r is the particle-hole separation. We see that this dipole moment is independent of chain length.

Table 8.1 shows that the dipole moment is largest for close lying exciton states, as the integral in eqn (8.31) is maximized when $|n_p - n_q| = 1$. It also shows that the dipole moment increases as n_p and n_q increase, as the effective-particle wavefunction, $\psi_n(r)$, spreads out (or the particle-hole separation increases) as n increases (as shown in Appendix E).

8.3.4.2   The strong-coupling (Mott-Hubbard) limit

- For transitions between the ground state and excited states the transition dipole moments are $\sim \sqrt{t/U} \approx 0$. However, the oscillator strength to the lowest optically allowed state is $O(L)$, in order to satisfy the oscillator sum rule.

- For transitions between two excited states, $|p\rangle$ and $|q\rangle$, the transition dipole moment is given by eqn (8.31), with the same selection rules as in the weak-coupling limit.

8.4   Nonlinear optical processes

In a centro-symmetric molecule the polarization must reverse sign under a reversal of the electric field. This means that all even powers of the electric field vanish in eqn (8.1). It therefore follows that in a centro-symmetric molecule $\chi^{(n)}$ vanishes for all the even ns. Since conjugated polymers are often centro-symmetric,

Table 8.1 *Matrix elements, $\langle n|\hat{x}|m\rangle = 1/2a \int dr\psi_n(r)r\psi_m(r)$, (in units of the repeat distance, $2a$) using the weak-coupling exciton theory ($U = 10$ eV, $t = 2.5$ eV, and $\delta = 0.2$)*

n	m	$\langle n|\hat{x}|m\rangle$	n	m	$\langle n|\hat{x}|m\rangle$
1	2	0.914	3	4	3.075
1	3	0	3	5	0
1	4	0.030	3	6	0.293
1	5	0	4	5	4.137
1	6	0.005	4	6	0
2	3	1.467	5	6	6.626
2	4	0			
2	5	0.271			
2	6	0			

the lowest nonzero nonlinear susceptibility in conjugated polymers is therefore usually $\chi^{(3)}$.

The general expression for the third order nonlinear susceptibility is

$$\chi^{(3)}_{\alpha\beta\gamma\delta}(-\omega_\sigma;\omega_1,\omega_2,\omega_3) = \frac{N}{3!\epsilon_0\hbar^3}S\sum_{LMN}$$

$$[\frac{\langle 0|\hat{\mu}_\alpha|L\rangle\langle L|\hat{\mu}_\beta|M\rangle\langle M|\hat{\mu}_\gamma|N\rangle\langle N|\hat{\mu}_\delta|0\rangle}{(\Omega_L - \omega_\sigma)(\Omega_M - \omega_2 - \omega_1)(\Omega_N - \omega_3)} + \frac{\langle 0|\hat{\mu}_\beta|L\rangle\langle L|\hat{\mu}_\gamma|M\rangle\langle M|\hat{\mu}_\delta|N\rangle\langle N|\hat{\mu}_\alpha|0\rangle}{(\Omega_L + \omega_1)(\Omega_M - \omega_2 - \omega_3)(\Omega_N - \omega_3)}$$

$$+ \frac{\langle 0|\hat{\mu}_\gamma|L\rangle\langle L|\hat{\mu}_\delta|M\rangle\langle M|\hat{\mu}_\alpha|N\rangle\langle N|\hat{\mu}_\beta|0\rangle}{(\Omega_L + \omega_1)(\Omega_M + \omega_1 + \omega_2)(\Omega_N - \omega_3)} + \frac{\langle 0|\hat{\mu}_\delta|L\rangle\langle L|\hat{\mu}_\alpha|M\rangle\langle M|\hat{\mu}_\beta|N\rangle\langle N|\hat{\mu}_\gamma|0\rangle}{(\Omega_L + \omega_1)(\Omega_M + \omega_1 + \omega_2)(\Omega_N + \omega_\sigma)}]$$

$$(8.33)$$

where S represents the symmetrization operator that permutes the 3! pairs of $(\hat{\mu}_\alpha,\omega_1)$, $(\hat{\mu}_\beta,\omega_2)$, and $(\hat{\mu}_\gamma,\omega_3)$. As before, if the polymers are oriented along the x-axis, then the dominant susceptibility is $\chi^{(3)}_{xxxx}$.

8.4.1 *The essential states mechanism*

It is clearly a formidable task to evaluate eqn (8.33) for all the possible states. However, a considerable simplification occurs if we adopt the *essential states* mechanism, introduced by Mazumdar *et al.* (Dixit *et al.* 1991) following earlier work by Heflin *et al.* (1988) and Soos and Ramesesha (1989). The key idea behind this concept is that only a few states are strongly dipole connected, and that these states dominate the sum. In fact, as we saw in Section 8.3.4, the state with the largest dipole moment to the ground state is the lowest-lying exciton state, the $n = 1$ and $j = 1$, or the $1^1B_u^-$ state. This, in turn, is most strongly dipole connected to the nearest lying exciton state, namely, the $n = 2$ and $j = 1$, or the $m^1A_g^+$ state. Finally, the $m^1A_g^+$ state is connected to the $n = 3$ and $j = 1$, or the $n^1B_u^-$ state. Thus, the $1^1A_g^+$, $1^1B_u^-$, $m^1A_g^+$, and $n^1B_u^-$ states constitute the four essential states.

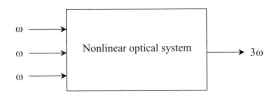

FIG. 8.6. Third harmonic generation process.

In the weak-coupling (Mott-Wannier) limit $m \equiv 2$ if particle-hole symmetry applies. Otherwise $m > 2$. In the strong-coupling (Mott-Hubbard) limit $m > 2$ always. The essentially states are shown schematically in Figs 6.7 and 6.9 for the weak-coupling and strong-coupling limits, respectively.

Usually, there is an unambiguous identification of the essential states. However, there are at least three reasons why this identification can become difficult. First, as noted in Table 8.1, the interstate dipole moments become larger for higher lying states. Thus, if there is a relatively large dipole moment between the ground state and a high lying $^{1}B_{u}^{-}$ state it is possible that another pathway significantly contributes to $\chi^{(3)}$. Second, it is possible that a state with a low principle quantum number, n, and a high pseudomomentum quantum number, j, is almost degenerate with a state with a high n and a low j. When this happens oscillator strength is transferred from the high n, low j state to the low n, high j state. Finally, as discussed in detail in Chapter 6, there may be two families of essential states, corresponding to there being both Mott-Wannier and Mott-Hubbard families of excitons. This scenario is most likely to happen in the intermediate-coupling regime.

Bearing in mind these caveats for the validity of the essential states mechanism, we shall now make the assumption that it is a reasonable approximation. This enables us to more readily interpret the third order nonlinear susceptibilities, and in particular, to relate experimental observations to the excited states of the polymer. We discuss this in the following sections.

8.4.2 *Third order harmonic generation*

In a third order harmonic generation process the system absorbs three photons of energy $\hbar\omega$ and emits one photon of energy $3\hbar\omega$, as shown schematically in Fig. 8.6. The susceptibility for this process is defined as

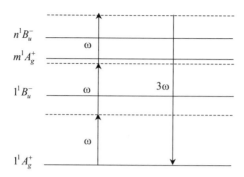

FIG. 8.7. Schematic energy level diagram, showing the energies of the $1^1B_u^-$, $m^1A_g^+$, and $n^1B_u^-$ states (bold), and the virtual transitions (dashed) for the dominant third harmonic generation process given by the first term on the right-hand side of eqn (8.34).

$$\chi^{(3)}(-3\omega; \omega, \omega, \omega) =$$

$$\frac{Ne^4}{\epsilon_0\hbar} \sum_{lp} \langle 1^1A_g^+|\hat{x}|l^1B_u^-\rangle\langle l^1B_u^-|\hat{x}|m^1A_g^+\rangle\langle m^1A_g^+|\hat{x}|p^1B_u^-\rangle\langle p^1B_u^-|\hat{x}|1^1A_g^+\rangle$$

$$[\frac{1}{(\Omega_{lB_u}-3\omega)(\Omega_{mA_g}-2\omega)(\Omega_{pB_u}-\omega)} + \frac{1}{(\Omega_{lB_u}+\omega)(\Omega_{mA_g}-2\omega)(\Omega_{pB_u}-\omega)}$$

$$+\frac{1}{(\Omega_{lB_u}+\omega)(\Omega_{mA_g}+2\omega)(\Omega_{pB_u}-\omega)} + \frac{1}{(\Omega_{lB_u}+\omega)(\Omega_{mA_g}+2\omega)(\Omega_{pB_u}+3\omega)}],$$

$$(8.34)$$

where the sum over l and p includes only the $1^1B_u^-$ and $n^1B_u^-$ states. (For clarity, we have also neglected the Franck-Condon factors in this expression.) The dominant term is the first one, represented by the energy level diagram shown in Fig. 8.7. There are one-photon resonances at $\hbar\omega = E(1^1B_u^-)$ and $\hbar\omega = E(n^1B_u^-)$, two-photon resonances at $\hbar\omega = E(m^1A_g^+)/2$, and three-photon resonances at $\hbar\omega = E(1^1B_u^-)/3$ and $\hbar\omega = E(n^1B_u^-)/3$. A comparison of the third order harmonic generation and the linear absorption is usually enough to allow an unambiguous identification of the $1^1B_u^-$, $m^1A_g^+$, and $n^1B_u^-$ states. Figure 8.8 shows a schematic sketch of the linear and third order harmonic generation coefficients corresponding to the energy level diagram of Fig. 8.7.

8.4.3   Electroabsorption

The electroabsorption is defined as the normalized change in transmission arising from a DC electric field, ξ,

$$-\frac{\Delta T}{T} = D\Delta\alpha, \qquad (8.35)$$

where D is the sample thickness and $\Delta\alpha$ is the change in the linear absorption. Since the electric field breaks the inversion symmetry, the $^1A_g^+$ and $^1B_u^-$ states

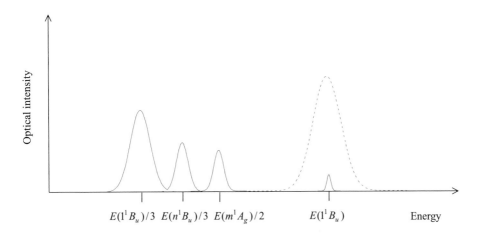

FIG. 8.8. Schematic plot of the third harmonic optical intensity (solid) and linear absorption (dashed).

are mixed, rendering the formerly disallowed $^1A_g^+$ states weakly allowed. $\Delta T/T$ may be calculated from the nonlinear third order susceptibilities via,

$$-\frac{\Delta T}{T} = \frac{\omega D \xi^2}{nc} \mathrm{Im}[\chi^{(3)}(-\omega;\omega,0,0)]. \tag{8.36}$$

In principle, electroabsorption experiments enable excitons to be distinguished from interband transitions, as excitons are subject to the *Stark effect*, while interband transitions are subject to the *Franz-Keldysh effect*. We now describe these two effects.

8.4.3.1 *The Stark effect* An electric field only affects the relative motion of an exciton and has no affect on the centre-of-mass motion. Thus, the total potential experienced by the electron-hole pair, $V_{\mathrm{tot}}(r)$, is

$$V_{\mathrm{tot}}(r) = V_{\mathrm{pot}}(r) - \frac{e^2}{4\pi\epsilon_0\epsilon|r|}, \tag{8.37}$$

where,

$$V_{\mathrm{pot}}(r) = e\xi r \tag{8.38}$$

arises from the electric field, ξ, and r is the relative coordinate. V_{tot} is sketched in Fig. 8.9. The maximum value of V_{tot} on the left-hand side is

$$V_m = -2e\sqrt{\frac{e\xi}{4\pi\epsilon_0\epsilon}}. \tag{8.39}$$

If the exciton binding energy is less that V_m the exciton immediately dissociates under the influence of the electric field. If the exciton binding energy is greater

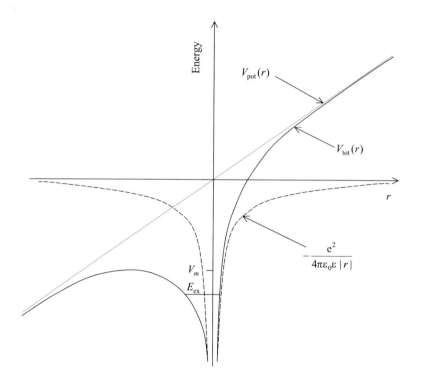

FIG. 8.9. The potential, $V_{\text{tot}}(r)$, as a function of the relative coordinate, r, experienced by an electron-hole pair in the presence of an electric field, ξ. V_m is the height of the barrier. The energy level of a bound exciton is denoted by E_{ex}. An exciton with a binding energy larger than V_m will decay in a finite time by the tunnelling of the hole through the potential barrier.

than V_m, however, there is a finite probability that the hole can tunnel through the potential barrier, and thus bound excitons acquire an electric-field dependent life-time.

The value of the electric field that immediately dissociates the exciton can also be estimated by a simple argument. The internal electric field in an exciton, ξ_{ex}, is

$$\xi_{\text{ex}} \sim \frac{E_{\text{ex}}}{e\langle r\rangle}, \tag{8.40}$$

where E_{ex} is the binding energy and $\langle r\rangle$ is the mean separation between the particle and hole. Now, in the hydrogenic model (see Appendix E, for example)

$$\langle r\rangle \sim \frac{e^2}{4\pi\epsilon_0\epsilon E_{\text{ex}}}. \tag{8.41}$$

Thus,

$$\xi_{ex} \sim \frac{4\pi\epsilon_0\epsilon E_{ex}^2}{e^3}. \qquad (8.42)$$

Setting $\xi = \xi_{ex}$ gives a condition for the dissociating electric field of the same order as that from eqn (8.39) when V_m is set to $-E_{ex}$.

To calculate the effects of an electric field it is necessary to add the term $V_{pot}(r)\psi_n(r)$ to the equation that describes the exciton wavefunction, $\psi_n(r)$ (namely, eqn (D.17)). For sufficiently small fields, the effect of $V_{pot}(r)$ on the exciton wavefunctions and energies can be calculated by perturbation theory. Now, since $V_{pot}(r)$ is an odd function of r and $\psi_n(r)$ are either even or odd functions of r it immediately follows that the first order corrections to the energy are zero. Thus, the change in energy to $|n\rangle$ to second order in perturbation theory is,

$$
\begin{aligned}
\Delta E_n &= \sum_{m\neq n} \frac{\langle n|V_{pot}(r)|m\rangle^2}{E_n - E_m} \\
&= e^2\xi^2 \frac{\langle n|\hat{r}|0\rangle^2}{E_n - E_0} + \frac{e^2\xi^2}{4a^2} \sum_{m>0, m\neq n} \frac{\left(\int \psi_n(r)r\psi_m(r)dr\right)^2}{E_n - E_m}. \qquad (8.43)
\end{aligned}
$$

The first term on the right-hand side arises from the dipole connection of $|n\rangle$ to the ground state, and is positive. The second term arises from the dipole connection of $|n\rangle$ to other excitons. As shown in Section 8.3.4, the matrix elements $\int \psi_n(r)r\psi_m(r)dr$ are largest when $m > n$. Thus, the sum is dominated by terms with negative denominators, and so this term is negative. The balance between the two terms on the right-hand side of eqn (8.43) determines whether the exciton is blue or red shifted by the electric field.

For the ground state we may write

$$
\begin{aligned}
\Delta E_0 &= -e^2\xi^2 \sum_{m>0} \frac{\langle m|\hat{r}|0\rangle^2}{\hbar\Omega_m} \\
&= -\frac{\alpha_M(0)\xi^2}{2}, \qquad (8.44)
\end{aligned}
$$

where,

$$\alpha_M(\omega) = \frac{e^2}{m} \sum_{m>0} \frac{f_m}{(\Omega_m^2 - \omega^2)} \qquad (8.45)$$

is the molecular polarizability, $\hbar\Omega_m = (E_m - E_0)$ and f_n is the oscillator strength. Equation (8.44) is the energy of an induced dipole in an electric field.

8.4.3.2 *The Franz-Keldysh effects* The effect of an electric field on an unbound particle-hole pair is nonperturbative, as the electron can gain an arbitrary amount of energy in the electric field by moving away from the hole. The effect of this is to reduce the band gap to zero. However, for there to be an optical transition the electron must tunnel away from the hole. This photon-induced tunnelling yields an optical absorption below the zero-field band edge as

$$\propto \exp\left(-A(E_g - \hbar\omega)^{3/2}\right), \tag{8.46}$$

where A is a material-dependent parameter.

There are also two other consequences of the electric field on the optical signature of an unbound particle-hole pair. First, the electroabsorption above the zero-field band gap exhibits oscillatory behaviour. This oscillatory behaviour can be traced to the oscillatory nature of the Airy functions, which are the solutions of the effective particle-hole equation in the absence of a Coulomb potential. Second, the position of the electroabsorption peaks vary as $\xi^{2/3}$, and not with the ξ^2 behaviour of excitons.

The Franz-Keldysh effects (Weiser and Horváth 1997) have been successfully used to distinguish the particle-hole continuum from exciton states in polydiacetylene crystals (Sebastian and Weiser 1981).

8.5   Size-dependencies of $\chi^{(n)}$

To conclude this chapter we discuss the size-dependencies of the electric susceptibilities. The weak-coupling exciton theory and the oscillator sum rule indicate that transition dipole moments from the ground state to an excited state are proportional to $\sqrt{L}$, whereas interexcited state transition dipole moments are independent of size. This result indicates that $\chi^{(1)}$ is a linear function of L for long chains.

$\chi^{(3)}$ processes that involve only two states are quadratic functions of L. However, most processes, such as that indicated in Fig. 8.7 which involve three or four states, will be linear functions of L. For short chains, however, there will be a supralinear dependence of $\chi^{(3)}$ on L, because, as shown in Fig. F.1, owing to confinement effects, the interexcited state transition dipole moments increase with chain lengths for lengths shorter than the root-mean-square particle-hole separations.

9

ELECTRONIC PROCESSES IN CONJUGATED POLYMERS

9.1 Introduction

A number of important electronic processes in conjugated polymers are introduced in this chapter. The emphasis is on describing electronic processes in single or weakly coupled polymers. We describe exciton (or energy) transfer, excited molecular complexes, charge transfer, and a theory of what determines the singlet exciton yield in light emitting polymers. Only in Section 9.4, where we describe screening of intramolecular excitations, are bulk systems considered. So, many electronic processes in bulk systems, such as the important topics of energy and electron transport, are not discussed. A discussion of electron transport in polymers may be found in Pope and Swenberg (1999) and the reviews by Rehwald and Kiess (1992), Bässler (2000), or Walker *et al.* (2002).

9.2 Exciton transfer

Energy is transferred from molecule to molecule (or more generally, from chromophore to chromophore) via the transfer of excitons. In this section we discuss resonant exciton transfer, whereby a 'donor' molecule in an excited state de-excites (usually to the ground state) while simultaneously an 'acceptor' molecule undergoes a transition to an excited state (usually from the ground state). This process in illustrated in Fig. 9.1.

The exciton transfer is coherent when the transfer time is fast compared to dissipative relaxation times, where the dissipative relaxation is usually inter or intramolecular vibrational relaxation and the transfer time is proportional to the inverse of the exciton transfer integral (described in the next section). In this limit the exciton is described by a wavefunction, whose time dynamics are controlled

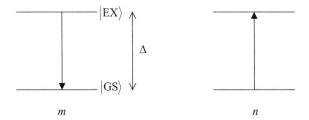

FIG. 9.1. Resonant exciton transfer between the donor molecule, m, initially in an excited state, $|EX\rangle$, and the acceptor molecule, n, initially in its ground state, $|GS\rangle$.

by the Schrödinger equation. Conversely, the exciton transfer is incoherent when the transfer time is slow compared to dissipative relaxation times. In this limit the exciton is described by a probability distribution, whose time dynamics are controlled by classical rate equations.

9.2.1   *Exciton transfer integral*

In this section we derive an expression for the transfer integral for the resonant exciton transfer depicted in Fig. 9.1. Consider a system consisting of two molecules, m and n. The total Hamiltonian for the system is the sum of the intramolecular Hamiltonians, H_m and H_n, and the intermolecular Hamiltonian, H_{mn}. The eigenstates of H_m and H_n are $\{|M\rangle_m\}$ and $\{|N\rangle_n\}$, respectively.

Suppose that initially molecule m is in an excited state, $|EX\rangle_m$, and molecule n is in its the ground state, $|GS\rangle_n$. In the absence of intermolecular interactions the initial state, $|I\rangle$, is a direct product of these molecular states:

$$|I\rangle = |EX\rangle_m |GS\rangle_n. \tag{9.1}$$

The transfer of energy results in a final state, $|F\rangle$, defined by

$$|F\rangle = |GS\rangle_m |EX\rangle_n. \tag{9.2}$$

This transfer of energy is mediated by the Coulomb interactions between the pair of molecules. Thus, the transfer integral, W_{mn} is defined as

$$W_{mn} = \langle F|H_{mn}^{e-e}|I\rangle, \tag{9.3}$$

where

$$H_{mn}^{e-e} = \sum_{\mathbf{r}\in m \mathbf{r}'\in n} \frac{e^2}{|\mathbf{r}-\mathbf{r}'|}, \tag{9.4}$$

and $\mathbf{r}$ and $\mathbf{r}'$ are the electronic coordinates.

It is convenient to separate W_{mn} into a component arising from the direct Coulomb interactions, denoted by J_{mn}, and another component arising from the exchange interactions, denoted by $\tilde{K}_{mn}$, namely,

$$W_{mn} = J_{mn} - \tilde{K}_{mn}. \tag{9.5}$$

The direct exciton transfer integral is responsible for the transfer of singlet excitons between pairs of molecules. This is the dominant process. The exchange exciton transfer integral is responsible for the transfer of triplet excitons between pairs of molecules. However, the exchange interactions decrease exponentially with distance, as they originate from the overlap of atomic wavefunctions, and consequently the exchange transfer integral is also very small. We therefore focus on the direct exciton transfer, J_{mn}.

Expressing the intermolecular direct Coulomb Hamiltonian in second quantization (as described in Section 2.6), namely,

$$H_{mn}^{e-e} = \sum_{i\in m j\in n} V_{ij}(\hat{N}_i - 1)(\hat{N}_j - 1),$$

(9.6)

and using eqns (9.1) and (9.2) we have that

$$
\begin{aligned}
J_{mn} &= \langle F| \left[\sum_{i\in m j\in n} V_{ij}(\hat{N}_i - 1)(\hat{N}_j - 1) \right] |I\rangle \\
&= \sum_{i\in m j\in n} V_{ij} \left[{}_m\langle GS|(\hat{N}_i - 1)|EX\rangle_m \right] \left[{}_n\langle EX|(\hat{N}_j - 1)|GS\rangle_n \right].
\end{aligned}
$$

(9.7)

When $|\mathbf{r}_i - \mathbf{r}_j|$ is large compared to the interatomic spacing the Coulomb potential is

$$V_{ij} = \frac{e^2}{|\mathbf{r}_i - \mathbf{r}_j|}.$$

(9.8)

Equation (9.7) shows why the direct Coulomb interaction only mediates singlet exciton transfer. Since the ground state is a singlet and since the operator $(\hat{N}_i - 1)$ preserves total spin, the excited state connected to the ground state in each of the square brackets must necessarily be a singlet.

We can see more clearly what J_{mn} represents by making the dipole approximation. We define $\tilde{\mathbf{r}}_i$ and $\tilde{\mathbf{r}}_j$ as the site coordinates relative to the centre-of-mass of their respective molecules,

$$
\begin{aligned}
\tilde{\mathbf{r}}_i &= \mathbf{r}_i - \mathbf{R}_m, \\
\tilde{\mathbf{r}}_j &= \mathbf{r}_j - \mathbf{R}_n,
\end{aligned}
$$

(9.9)

where $\mathbf{R}_m$ and $\mathbf{R}_n$ are the centre-of-mass coordinates of molecules m and n, respectively. Then, if $|\tilde{\mathbf{r}}_i - \tilde{\mathbf{r}}_j| << |\mathbf{R}_m - \mathbf{R}_n| \equiv |\mathbf{R}_{mn}|$ we can perform the dipole approximation and write,

$$\sum_{i\in m j\in n} \frac{1}{|\mathbf{r}_i - \mathbf{r}_j|} \approx \frac{\sum_{i\in m} \tilde{\mathbf{r}}_i \cdot \sum_{j\in n} \tilde{\mathbf{r}}_j}{|\mathbf{R}_{mn}|^3} - \frac{3\left(\sum_{i\in m} \mathbf{R}_{mn} \cdot \tilde{\mathbf{r}}_i\right)\left(\sum_{j\in n} \mathbf{R}_{mn} \cdot \tilde{\mathbf{r}}_j\right)}{|\mathbf{R}_{mn}|^5}.$$

(9.10)

Then, substituting into eqn (9.7)

$$J_{mn} = \kappa_{mn} J_{mn}^0,$$

(9.11)

where

$$J_{mn}^0 = \frac{[{}_m\langle GS|\hat{\mu}_m|EX\rangle_m][{}_n\langle EX|\hat{\mu}_n|GS\rangle_n]}{|\mathbf{R}_{mn}|^3},$$

(9.12)

and

$$\kappa_{mn} = \hat{\mathbf{r}}_m \cdot \hat{\mathbf{r}}_n - 3(\hat{\mathbf{R}}_{mn} \cdot \hat{\mathbf{r}}_m)(\hat{\mathbf{R}}_{mn} \cdot \hat{\mathbf{r}}_n), \tag{9.13}$$

is an orientational factor. $\hat{\mathbf{r}}_m$ and $\hat{\mathbf{R}}_{mn}$ are the unit vector parallels to $\hat{\mu}_m$ and $\mathbf{R}_{mn}$, respectively. $\hat{\mu}_m$ is the electronic dipole operator for molecule m, defined by

$$\hat{\mu}_m = e \sum_{i \in m} \tilde{\mathbf{r}}_i (\hat{N}_i - 1). \tag{9.14}$$

In the dipole approximation we therefore see that J_{mn} represents the interaction between a transition dipole moment on molecule m with a transition dipole moment on molecule n. Since these matrix elements are only nonzero for dipole-allowed transitions this level of approximation describes the transfer of dipole-allowed singlet excitons. For centrosymmetric molecules these will be the ^1B_u excitons. Higher multipole expansions of the Coulomb potential will describe the transfer of dipole-forbidden singlet excitons.

The dipole approximation is valid provided that the spatial extent of the transition dipole moments are much smaller than the separation between the centre-of-masses of the pair of molecules. However, since the transition dipole moments for polymers are large (and scale as $\sqrt{L}$ for transitions from the ground state), this assumption is not necessarily valid.

To conclude this section we note that the triplet exciton transfer integral is

$$
\begin{aligned}
\tilde{K}_{mn} &= \langle F| \left[\sum_{i \in m j \in n} \sum_{\sigma} K_{ij} c_{i\sigma}^{\dagger} c_{i\bar{\sigma}} c_{j\bar{\sigma}}^{\dagger} c_{j\sigma} \right] |I\rangle \\
&= \sum_{i \in m j \in n} \sum_{\sigma} K_{ij} \left[{}_m\langle GS| c_{i\sigma}^{\dagger} c_{i\bar{\sigma}} |EX\rangle_m \right] \left[{}_n\langle EX| c_{j\bar{\sigma}}^{\dagger} c_{j\sigma} |GS\rangle_n \right], \tag{9.15}
\end{aligned}
$$

with K_{ij} defined in eqn (2.30).

9.2.2 Coherent transfer

In this section we discuss the coherent motion of excitons through an assembly or aggregate of N molecules. This is applicable when $\hbar/J$ is much smaller than the dissipative relaxation times.

Suppose that the operator $E_m^{\dagger}$ creates an exciton on the mth molecule. Thus, it corresponds to exciting the molecule from its ground state to the state $|EX\rangle_m$ with an excitation energy Δ. The Hamiltonian that describes the coherent motion of the exciton through the aggregate is

$$H = \sum_{mn} \left(J_{mn} E_m^{\dagger} E_n + J_{nm} E_n^{\dagger} E_m \right) + \Delta \sum_m E_m^{\dagger} E_m. \tag{9.16}$$

For simplicity we will assume that only one exciton is excited in the aggregate. Then the basis states are 'single-particle' basis states of the form,

$$|m\rangle = E_m^{\dagger} \prod_{n=1}^{N} |GS\rangle_n. \tag{9.17}$$

(a)   (b)

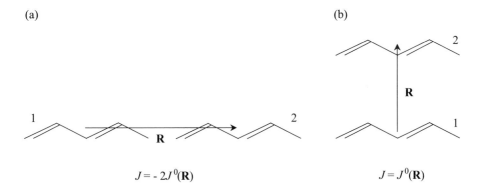

$$J = -2J^0(\mathbf{R})$$ $$J = J^0(\mathbf{R})$$

FIG. 9.2. Collinear (a) and parallel (b) arrangement of a pair of molecules in a dimer. The exciton transfer integral, J, is negative for the collinear arrangement and positive for the parallel arrangement.

$|m\rangle$ represents an exciton localized on the mth molecule. The eigensolutions of eqn (9.16) were discussed in detail in Chapter 3. (See also Appendix A.) Here we will review those solutions and discuss the particularly important results for two coupled molecules, namely a dimer.

9.2.2.1   *Dimers*   We discuss dimers for two particular geometrical arrangements, namely *collinear* and *parallel*, as shown in Fig. 9.2. In the collinear arrangement the exciton transfer integral, $J = -2J^0(R)$, whereas in the parallel arrangement, $J = J^0(R)$, where $J^0(R)$ is defined in eqn (9.12). In both cases there are bonding, $|+\rangle$, and antibonding, $|-\rangle$, solutions:

$$|+\rangle = \frac{1}{\sqrt{2}}(|1\rangle + |2\rangle) \tag{9.18}$$

and

$$|-\rangle = \frac{1}{\sqrt{2}}(|1\rangle - |2\rangle) \tag{9.19}$$

with energies $(\Delta + J)$ and $(\Delta - J)$, respectively. $|m\rangle$ is defined in eqn (9.17). Thus, in the collinear arrangement the bonding solution is the lower energy state, whereas in the parallel arrangement the reverse is true.

The transition dipole moment from the ground state to the excited states of the dimer is

$$\langle \mathrm{GS}|\hat{\mu}_T|\pm\rangle = \langle \mathrm{GS}|\,(\hat{\mu}_1 + \hat{\mu}_2)\,\frac{1}{\sqrt{2}}\,(|1\rangle \pm |2\rangle)\,, \tag{9.20}$$

where $|\mathrm{GS}\rangle = |\mathrm{GS}\rangle_1|\mathrm{GS}\rangle_2$. Evaluating this expression for equivalent molecules gives

$$\langle \mathrm{GS}|\hat{\mu}_T|+\rangle = \sqrt{2}\langle \hat{\mu}^{GE}\rangle \tag{9.21}$$

and

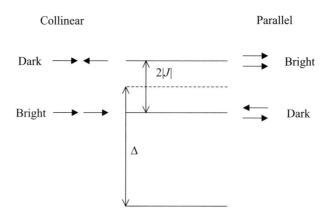

FIG. 9.3. The energy level diagram and schematic representation of the dipoles in the collinear and parallel arrangement of a pair of molecules in a dimer.

$$\langle \text{GS}|\hat{\mu}_T|-\rangle = 0, \tag{9.22}$$

where $\langle \hat{\mu}^{GE} \rangle$ is the transition dipole moment for a single molecule.

Equations (9.21) and (9.22) tell us that there is constructive interference of the dipole moments of the two molecules in the bonding state. Conversely, there is destructive interference in the antibonding state. Thus, the collinear arrangement is 'bright' and red-shifted, while the parallel arrangement is 'dark' with the dipole-active state blue-shifted. The splitting between these two states, $2J$, is known as the Davydov splitting. The energy level diagram and a schematic representation of the dipoles in the dimer are illustrated in Fig. 9.3.

9.2.2.2 *Aggregates* The analysis of the previous section is easily extended to aggregates of molecules. An aggregate of collinear molecules is known as a J-aggregate, while an aggregate of parallel molecules is known as an H-aggregate. Assuming only nearest-neighbour transfer the energies of the delocalized exciton form a band with energies,

$$\epsilon_\beta = \Delta + 2J \cos(\beta R). \tag{9.23}$$

The corresponding eigenstates are

$$|\beta\rangle = \sqrt{\frac{2}{N+1}} \sum_m \sin(\beta mR)|m\rangle, \tag{9.24}$$

where $\beta = j\pi/(N+1)R$, j satisfies $1 \leq j \leq N$ and R is the distance between molecules. The exciton, or Davydov, band width is $|4J|$.

As for the case of dimers, there is interference between the transition dipole moments of each molecule in the aggregate. Extending the derivation of the previous section, we have that

$$\langle \text{GS}|\hat{\mu}_T|\beta\rangle = 0, \text{ for even } j, \tag{9.25}$$

and

$$\langle \text{GS}|\hat{\mu}_T|\beta\rangle \approx \frac{\sqrt{N}\langle\hat{\mu}^{GE}\rangle}{j}, \text{ for odd } j, \tag{9.26}$$

(in analogy to eqns (F.10) and (F.11)). Again, $\langle\hat{\mu}^{GE}\rangle$ is the transition dipole moment of a single molecule. Notice that the total transition dipole moment scales as $\sqrt{N}$ in eqn (9.26), and thus the oscillator strength scales as N, as it must do to satisfy the oscillator sum rule, eqn (8.7).

This interference means that the lowest excited state of a J-aggregate (namely, $j = 1$) is *superluminescent*, whereas for an H-aggregate it is dark.

9.2.2.3 *Quantum mechanical dynamics*   The dynamics in the coherent regime are determined by the time-dependent Schrödinger equation,

$$i\hbar\frac{\partial|\Psi(t)\rangle}{\partial t} = H|\Psi(t)\rangle. \tag{9.27}$$

For the exciton transfer Hamiltonian defined by eqn (9.16) the general time dependent state is of the form,

$$|\Psi(t)\rangle = \sum_m a_m(t)|m\rangle, \tag{9.28}$$

where the basis states $\{|m\rangle\}$ are defined by eqn (9.17).

If the pth molecule (or chromophore) of the aggregate is excited at time $t = 0$, so that the exciton is initially localized on the pth molecule and $|\Psi(0)\rangle = |p\rangle$, the subsequent dynamics are given by,

$$a_m(t) = \left(\frac{2}{N+1}\right)\sum_\beta \exp(-i\epsilon_\beta t/\hbar)\sin(\beta mR)\sin(\beta pR), \tag{9.29}$$

where ϵ_β is given by eqn (9.23). The probability that the exciton is localized on the mth molecule at a time t is then $|a_m(t)|^2$.

For the dimer this result gives the characteristic quantum mechanical oscillatory behaviour. If molecule 1 is excited at time $t = 0$, the subsequent probabilities of finding the exciton on molecules 1 and 2 are,

$$P_1(t) = |a_1(t)|^2 = \cos^2(Jt/\hbar) \tag{9.30}$$

and

$$P_2(t) = |a_2(t)|^2 = \sin^2(Jt/\hbar), \tag{9.31}$$

respectively, with a period h/J. As we shall see in Section 9.2.4, dissipative phenomena dampen these oscillations.

9.2.3 *Incoherent transfer*

When the dissipative relaxation time is shorter than the exciton transfer time phase coherence of the exciton wavefunction can no longer be maintained and a probabilistic interpretation is more appropriate.

The probability, $P_m(t)$, that the exciton is on the mth molecule at time t is determined by the rate equation,

$$\frac{\partial P_m(t)}{\partial t} = -\sum_n \left(k_{mn} P_m(t) - k_{nm} P_n(t) \right), \qquad (9.32)$$

where the rates, k_{mn}, are usually determined by the Fermi Golden Rule. The rate for dipole-allowed singlet exciton transfer via the direct Coulomb interaction was determined by Förster (1951), and is known as Förster transfer. The rates for dipole-forbidden singlet or triplet exciton transfer were determined by Dexter (1953), and is known as Dexter transfer. We describe both of these processes in the next two sections.

Equation (9.32) has been widely used to model exciton transport in conjugated materials. See Movaghar *et al.* (1986) and Meskers *et al.* (2001) for an example of this approach. In Section 9.2.4 we briefly describe the more general density matrix formalism to model exciton transport.

9.2.3.1 *Förster transfer* The rate for direct exciton transfer between a donor molecule, D, and an acceptor molecule, A, is determined by the Fermi Golden Rule expression,

$$k_{DA} = \frac{2\pi}{\hbar} |J_{DA}|^2 \delta(E_F - E_I), \qquad (9.33)$$

where the δ-function ensures energy conservation. $|J_{DA}|^2$ is determined by eqn (9.7) (with the m and n replaced by D and A), and E_I and E_F are the initial and final energies, respectively.

Using the identity that,

$$\delta(x + y) = \int_{-\infty}^{\infty} \delta(x - z)\delta(y + z)dz, \qquad (9.34)$$

and noting that $E_I = E_F = \Delta$, where Δ is the exciton energy, as defined in Fig. 9.1, we rewrite eqn (9.33) as,

$$k_{DA} = \frac{2\pi}{\hbar} |J_{DA}|^2 \int_{-\infty}^{\infty} \delta(\Delta - \hbar\omega)\delta(\hbar\omega - \Delta)d(\hbar\omega). \qquad (9.35)$$

We interpret the first delta function as representing the absorption of energy by the acceptor molecule, while the second delta function represents the emission of energy by the donor.

In the dipole approximation we have,

$$k_{DA} = \frac{2\pi}{\hbar} \frac{\kappa_{DA}^2 |\hat{\mu}_D^{GE}|^2 |\hat{\mu}_A^{GE}|^2}{|\mathbf{R}_{DA}|^6} \int_{-\infty}^{\infty} \delta(\Delta - \hbar\omega)\delta(\hbar\omega - \Delta)d(\hbar\omega), \qquad (9.36)$$

where we have used eqns (9.11) - (9.13). Thus, using the expressions for the donor emission and acceptor absorption (eqn (8.9)) spectra,

$$I_D(\omega) = \frac{4\hbar\omega^3 |\hat{\mu}_D^{GE}|^2}{3n^3 c^3} \delta(\Delta - \hbar\omega) \qquad (9.37)$$

and

$$\alpha_A(\omega) = \frac{\pi\omega N |\hat{\mu}_A^{GE}|^2}{3nc\epsilon_0} \delta(\hbar\omega - \Delta), \qquad (9.38)$$

respectively,[37] the Förster transfer rate can be expressed as an overlap of these spectral functions:

$$k_{DA} = \frac{9n^4 c^4 \epsilon_0 \kappa_{DA}^2}{2N|\mathbf{R}_{DA}|^6} \int_0^{\infty} \frac{I_D(\omega)\alpha_A(\omega)}{\omega^4} d\omega. \qquad (9.39)$$

Here, n is the refractive index of the medium, c is the speed of light in vacuo and N is the number of polymers per unit volume. This formulation encapsulates the physical process of resonant exciton transfer, namely simultaneous emission of energy by the donor and absorption of energy by the acceptor in an energy conserving process. Notice, however, that a physical photon is not exchanged between the donor and acceptor in this process.

We also notice that, unlike the case of coherent transfer - where the transfer rate is $\propto |J_{DA}|$ and hence $|\mathbf{R}_{DA}|^{-3}$ - for incoherent transfer the rate is $\propto |J_{DA}|^2$ and hence $|\mathbf{R}_{DA}|^{-6}$. It is customary to define a Förster radius, R_F, at which the Förster transfer rate is equal to the radiative emission rate, $k_R = 1/\tau_R$. Then,

$$k_{DA} = \frac{1}{\tau_R} \left(\frac{R_F}{|\mathbf{R}_{DA}|}\right)^6. \qquad (9.40)$$

Thus, for molecules separated by distances less than R_F exciton transfer is more likely to occur than radiative recombination.

9.2.3.2 *Dexter transfer* A similar expression determines the transfer rate for triplet excitons, namely,

$$k_{DA} = \frac{2\pi}{\hbar} |\tilde{K}_{DA}|^2 \int F_D(E)F_A(E)dE, \qquad (9.41)$$

where $\tilde{K}_{DA}$ is given by eqn (9.15). The spectral functions are defined by,

---

[37] The factors of 1/3 arise from the orientational average.

$$F_D(E) = \frac{E^{-3}I_D(E)}{\int E^{-3}I_D(E)dE}, \tag{9.42}$$

where $I_D(E)$ is the emission curve of the donor and

$$F_A(E) = \frac{E^{-1}\sigma_A(E)}{\int E^{-1}\sigma_A(E)dE}, \tag{9.43}$$

where $\sigma_A(E)$ is the absorption curve of the acceptor.

9.2.4 The density matrix approach

In general the exciton dynamics exhibits both coherent and incoherent behaviour, where the incoherence arises from the coupling of the system to a dissipative environment. This is conveniently modelled by an equation of motion for the reduced density operator, $\hat{\rho}$, defined by

$$\hat{\rho} = Tr\{\hat{W}(t)\}, \tag{9.44}$$

where $\hat{W}(t)$ is the full density operator and the trace is over the degrees of freedom of the environment. (See May and Kühn (2000) for more details of this approach.)

In the localized exciton basis $\{|m\rangle\}$, defined by eqn (9.17), the matrix elements of the reduced density operator are

$$\rho_{mn} = \langle m|\hat{\rho}|n\rangle. \tag{9.45}$$

The diagonal elements, ρ_{nn}, represent the classical populations, P_n, while the off-diagonal elements, ρ_{mn}, describe the coherences between the quantum states $|m\rangle$ and $|n\rangle$.

The equation of motion for the matrix elements is

$$\frac{\partial \rho_{mn}}{\partial t} = -i\omega_{mn}\rho_{mn} - \frac{i}{\hbar}\sum_{\ell}(J_{m\ell}\rho_{\ell n} - J_{\ell n}\rho_{m\ell})$$
$$-\delta_{mn}\sum_{\ell}(k_{m\ell}\rho_{mm} - k_{\ell m}\rho_{\ell\ell}) - (1 - \delta_{mn})(\gamma_m + \gamma_n)\rho_{mn}, \tag{9.46}$$

where $\omega_{mn} = (\Delta_m - \Delta_n)/\hbar$. The first two terms on the right-hand side represent the coherent dynamics described in Section 9.2.2.3. The final two terms are the dissipative terms: the third term represents the incoherent dynamics described by the rate equation, eqn (9.32), while the final term represents coherence dephasing.

For a system in contact with a thermal heat bath equilibrium is achieved provided that detailed balance is satisfied, namely

$$\frac{k_{m\ell}}{k_{\ell m}} = \exp(-\hbar\omega_{\ell m}/k_BT). \tag{9.47}$$

The competition between coherence and incoherence is readily illustrated by the dimer, which serves as a model of a two-level system. If molecule 1 is initially excited at time $t = 0$, the subsequent populations of molecules 1 and 2 are,

$$P_1(t) = \rho_{11}(t) = \frac{1}{2}\left(1 + \exp(-\alpha t)\cos(\omega t)\right) \tag{9.48}$$

and

$$P_2(t) = \rho_{22}(t) = \frac{1}{2}\left(1 - \exp(-\alpha t)\cos(\omega t)\right), \tag{9.49}$$

respectively. The parameters are,

$$\alpha = k + \gamma \tag{9.50}$$

and

$$\omega = \left(\frac{4J^2}{\hbar^2} - (k - \gamma)^2\right)^{1/2}, \tag{9.51}$$

where $J = J_{12} = J_{21}$, $k = k_{12} = k_{21}$, and $\gamma = \gamma_1 = \gamma_2$.

Equations (9.48) and (9.49) indicate that the system exhibits the 'classical' populations after a relaxation (or decoherence) time, α^{-1}. Evidently, we also see that the system is overdamped when $2|J|/\hbar < |k - \gamma|$.

9.3 Excited molecular complexes

9.3.1 Excimers

Excited dimers, or excimers, were introduced in Section 9.2.2.1 in the context of resonant exciton transfer between two coupled molecules. See (Gordon and Ware 1975) for a review. This resonant transfer results in the 'bonding' and 'antibonding' states, described by eqns (9.18) and (9.19), respectively.[38] Thus, the molecules in the lower excited state of a dimer experience an attraction - mediated by the exchange of excitons - that is not present in the ground state.

As well as the intramolecular exciton component, excimers also have a charge-transfer component, corresponding to an electron on one of the molecules and a hole on the other. Representing the component with a hole on molecule m and an electron on molecule n as $|+\rangle_m|-\rangle_n$, an excimer state is written as

$$|\Psi\rangle = c_1\left(|EX\rangle_m|GS\rangle_n \pm |GS\rangle_m|EX\rangle_n\right) + c_2\left(|+\rangle_m|-\rangle_n \pm |-\rangle_m|+\rangle_n\right). \tag{9.52}$$

The charge-transfer component further stabilizes the excimer in two ways. First, the intermolecular Coulomb interaction induces an attraction between the electron-hole pair, resulting in a weakly bound charge-transfer exciton. Second, the inter-chain one-electron Hamiltonian couples the charge-transfer component with the exciton component.

A schematic illustration of the adiabatic potential energy surfaces for the dimer is shown in Fig. 9.4. Since the pair of molecules in the lowest excited state

---

[38]Note, however, that for the parallel arrangement the 'antibonding' state has lower energy.

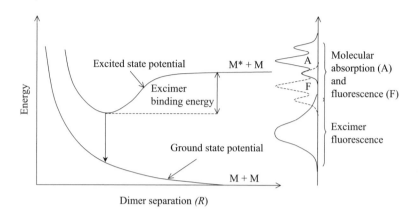

FIG. 9.4. Schematic illustration of the adiabatic potential energy surfaces for the ground state and lowest excited state of a dimer as a function of the molecular separation, R. Also illustrated are the excimer fluorescence spectrum and the absorption and emission spectra for a single molecule.

experience an attraction that is not present in the ground state the equilibrium geometry of the excimer corresponds to a repulsive ground state potential. Thus the excimer emission is broad, featureless and red-shifted in comparison to the molecular emission. Notice that there is also no associated absorption from the ground state to the excimer.

Equation (9.52) indicates that there are four components to the excimer state, $|\Psi\rangle$. Correspondingly, there are four excimer eigenstates. The lowest two in energy are predominately composed of the bonding or antibonding exciton wavefunctions, already described in Section 9.2.2.1. These are split in energy by roughly $2J$ and centred around the exciton energy, Δ, as illustrated in Fig. 9.3. The highest two eigenstates in energy are predominately composed of the bonding or antibonding charge-transfer wavefunctions. These are split in energy by roughly the intermolecular transfer integral and centred around the charge-transfer energy. This ordering of the 'excimer eigenstates' is quite different from an exciplex, as we describe below, where a particular charge-transfer component is usually the lowest state.

9.3.2 Exciplexes

An exciplex is an *excited* com*plex* of two different molecules. Thus an exciplex state is represented as

$$|\Psi\rangle = c_1|\text{EX}\rangle_m|\text{GS}\rangle_n + c_2|\text{GS}\rangle_m|\text{EX}\rangle_n + c_3|+\rangle_m|-\rangle_n + c_4|-\rangle_m|+\rangle_n. \quad (9.53)$$

In general, for molecules with similar HOMO-LUMO gaps the exciplex will be dominated by the charge-transfer component $|+\rangle_D|-\rangle_A$, where D and A stand for donor and acceptor, respectively. The donor and acceptor are defined such

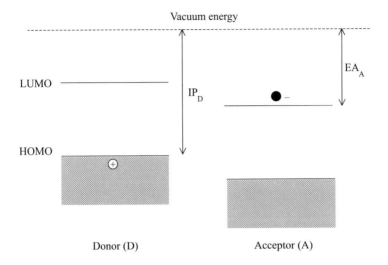

FIG. 9.5. The charge-transfer state of an exciplex, with the ionization potential, IP_D, and the electron affinity, EA_A, of the donor (D) and acceptor (A) also shown.

that $IP_D < IP_A$ and $EA_A > EA_D$, where IP is the ionization potential (defined as the energy required to remove an electron) and EA is the electron affinity (defined as the energy gained by accepting an electron), as illustrated in Fig. 9.5. Then, neglecting exciton and charge transfer terms, the lowest excited state energy, Δ is

$$\Delta = IP_D - EA_A - C(R), \qquad (9.54)$$

where $-C(R)$ is the Coulomb potential energy between the electron-hole pair.

The transition dipole moment of excimers and exciplexes with the ground state (and hence their life-times) is a function of the relative orientation of the two molecules. As we have seen in Section 9.2.2.1, parallel molecules in an excimer have a zero transition dipole moment, whereas collinear molecules have an enhanced dipole moment.

9.4   Screening of intramolecular states

We described in Section 9.2 how excitons transfer from chain to chain as a result of interchain interactions. Another important consequence of interchain interactions is that they screen *intrachain* excitations. An understanding of the strength of this screening is essential to quantitatively predict the energies of intrachain excitations of polymers in a bulk environment. In this section we describe how these screening energies can be estimated using second order perturbation theory, and we show why screening of excitons in polymers is quite different from that in conventional semiconductors.

Consider an assembly of N molecules described by the Hamiltonian,

$$H = H^{\text{intra}} + H^{\text{inter}}. \tag{9.55}$$

The intermolecular Hamiltonian, H^{inter}, is regarded as the perturbation on the sum of intramolecular Hamiltonians,

$$H^{\text{intra}} = \sum_n H_n^{\text{intra}}, \tag{9.56}$$

where H_n^{intra} is the intramolecular Hamiltonian for the nth molecule. H^{inter} is a sum of pair-wise interactions between pairs of molecules:

$$H^{\text{inter}} = \frac{1}{2} \sum_{m \neq n} H_{mn}^{e-e} \tag{9.57}$$

and H_{mn}^{e-e} is defined in eqn (9.6) (as we are neglecting electron-transfer processes). A general eigenstate of H^{intra} is a direct product of intramolecular eigenstates,

$$|A\rangle = \prod_m |M\rangle_m. \tag{9.58}$$

To second order in perturbation theory the change in energy arising from the intermolecular interactions is

$$\Delta E_A = \langle A|H^{\text{inter}}|A\rangle + \sum_{A \neq A'} \frac{\langle A'|H^{\text{inter}}|A\rangle^2}{E_A - E_{A'}}. \tag{9.59}$$

Consider the general matrix element, $\langle A'|H^{\text{inter}}|A\rangle$. Using the fact that H^{inter} is a sum over pairwise interactions this matrix element is

$$\langle A'|H^{\text{inter}}|A\rangle = \frac{1}{2} \sum_{m \neq n} \langle A'|H_{mn}^{e-e}|A\rangle. \tag{9.60}$$

Further, since H_{mn}^{e-e} only operates on molecules m and n we have,

$$\langle A'|H_{mn}^{e-e}|A\rangle = \sum_{i \in m, j \in n} V_{ij} \left[{}_m\langle M'|(\hat{N}_i - 1)|M\rangle_m \right] \left[{}_n\langle N'|(\hat{N}_j - 1)|N\rangle_n \right]. \tag{9.61}$$

For neutral systems with particle-hole symmetry the diagonal terms are zero, because $\langle \hat{N} - 1 \rangle = 0$ (as shown in Appendix B). Thus, the first order corrections to the energy are zero, and we need to consider the off-diagonal matrix elements to second order in perturbation theory, where both $|M'\rangle_m \neq |M\rangle_m$ and $|N'\rangle_n \neq |N\rangle_n$.

In general, therefore, the sum over A' in eqn (9.59) is a sum over all pairs of molecules m and n and all states $|M'\rangle_m$ and $|N'\rangle_n$ connected to $|M\rangle_m$ and

$|N\rangle_n$, respectively by H_{mn}^{e-e}. There are two sets of terms to consider for identical molecules. First, there are terms from the degenerate states associated with the transfer of energy from molecule to molecule, namely $|M'\rangle_m = |N\rangle_m$ and $|N'\rangle_n = |M\rangle_n$. These are the resonant terms already considered in Section 9.2. Second, there are the nonresonant terms that lead to the shift in state energies. We now consider these latter terms.

Since our purpose is to calculate the screening of an excitation in a general molecule by the other molecules, let us now consider the situation where molecule p will be in a general state $|P\rangle_p$ and the remaining $\{q\}$ molecules will be in their ground states, $\{|\text{GS}\rangle_q\}$. The sum over A' in eqn (9.59) is now over all molecules q and all the connected eigenstates. In general this summation is not practical. However, the essential physics underlying the screening is revealed when we make two simplifying assumptions. First, for sufficiently distant molecules we can again invoke the dipole approximation. Suppose that,

$$|A\rangle = |P\rangle_p \prod_q |\text{GS}\rangle_q \qquad (9.62)$$

and

$$|A'\rangle = |P'\rangle_p |Q\rangle_q \prod_{q' \neq q} |\text{GS}\rangle_{q'}, \qquad (9.63)$$

then,

$$\langle A'|H^{\text{inter}}|A\rangle = \frac{\kappa_{pq}}{|\mathbf{R}_{pq}|^3} \left[_p\langle P'|\hat{\mu}_p|P\rangle_p \right] \left[_q\langle Q|\hat{\mu}_q|\text{GS}\rangle_q \right]. \qquad (9.64)$$

Second, we invoke the 'essential-states' mechanism, described in Chapter 8, which states that only a few states are strongly dipole connected. These are the ground state and the lowest (pseudo)momentum eigenstate of each family of exciton states. For centro-symmetric polymers these states are:

$$1^1A_g \leftrightarrow 1^1B_u \leftrightarrow m^1A_g \leftrightarrow n^1B_u, \text{ etc.} \qquad (9.65)$$

Using these two approximations, we now derive simplified expressions for the screening of the ground state and some of the key low-lying states.

1. *The ground state* of the system corresponds to all molecules being in their ground state. The dominant term in the sum will be to the state with all pairs of molecules p and q being in their lowest excitation, the 1^1B_u state. Thus,

$$\Delta E_{1A} = -\frac{\left| \sum_q \frac{\kappa_{pq}}{|\mathbf{R}_{pq}|^3} \langle 1^1B_u|\hat{\mu}|1^1A_g\rangle^2 \right|^2}{2\Delta E(1^1B_u)}, \qquad (9.66)$$

where $\Delta E(X)$ is the transition energy of the state X.

2. *The first excited state* of the system corresponding to molecule p in the lowest excited state (1^1B_u) and the rest in their ground states. Again, the dominant term in the sum will be to the state with next highest excitation in all pairs of molecules p and q. Thus,

$$\Delta E_{1B} = -\frac{\left|\sum_q \frac{\kappa_{pq}}{|\mathbf{R}_{pq}|^3}\langle 1^1B_u|\hat{\mu}|1^1A_g\rangle\langle m^1A_g|\hat{\mu}|1^1B_u\rangle\right|^2}{\Delta E(m^1A_g)}. \tag{9.67}$$

Notice that the term corresponding to the de-excitation of molecule p and the excitation of another molecule to the 1^1B_u state is a resonant transfer, and therefore it is not included here.

3. *The second excited state* of the system corresponding to molecule p in the second excited state (m^1A_g) and the rest in their ground states. Now there are two important terms in the sum: one to the state with next highest excitation in all pairs of molecules p and q, and the other to a de-excitation in p and an excitation in q. Thus,

$$\Delta E_{mA} = -\frac{\left|\sum_q \frac{\kappa_{pq}}{|\mathbf{R}_{pq}|^3}\langle 1^1B_u|\hat{\mu}|1^1A_g\rangle\langle n^1B_u|\hat{\mu}|m^1A_g\rangle\right|^2}{\Delta E(n^1B_u) + \Delta E(1^1B_u) - \Delta E(m^1A_g)}$$

$$-\frac{\left|\sum_q \frac{\kappa_{pq}}{|\mathbf{R}_{pq}|^3}\langle 1^1B_u|\hat{\mu}|1^1A_g\rangle\langle 1^1B_u|\hat{\mu}|m^1A_g\rangle\right|^2}{2\Delta E(1^1B_u) - \Delta E(m^1A_g)}. \tag{9.68}$$

The expressions derived within the dipole and essential-states approximation indicate that:

- Screening is determined by induced transition dipole interactions, which decrease as R^{-6}. These are 'dispersive' or 'London' type interactions. The change in energy of the ground state given by eqn (9.66) is the van der Waal's interaction.
- Screening effectively reduces the energy of the intramolecular excitations. This screening is enhanced for the higher excited states, because the transition dipole moments between connected states increase (see Section 8.3.4) and the energy denominators decrease.
- The time-scale for the screening is determined by the time to establish a dipole in the dielectric, namely $\hbar/\Delta E(1^1B_u)$, and the time for the screened molecule to make a transition to a different state, namely $\hbar/(\Delta E(X_f) - \Delta E(X_i))$. As the excited state energies increase the energy differences decrease and this time scale diverges. Eventually this leads to a break-down in the perturbation theory, and to solvation-like screening, as in a point charge described below.

It is also instructive to estimate the screening of a free charge in the ground state, as the screening energy of an unbound particle-hole pair is twice this value.

Consider a doped particle on molecule p. Since in a doped molecule $\langle \hat{N} - 1 \rangle \neq 0$, the dominant matrix elements are of the type,

$$\sum_{i \in m j \in n} V_{ij} \left[{}_p\langle 1^1 A_g | (\hat{N}_i - 1)|1^1 A_g\rangle_p \right] \left[{}_q\langle 1^1 B_u | (\hat{N}_j - 1)|1^1 A_g\rangle_q \right]. \qquad (9.69)$$

This corresponds to the interaction of a monopole on molecule p with a transition dipole moment on molecule q, and represents the static solvation of the point charge.

As we have seen, the dielectric screening of the lowest excited states is dispersion-like, becoming solvation-like for the weakly bound states and charged states. Solvation-like screening occurs in inorganic semiconductors where the optical gap is much larger than the exciton binding energy. Thus, the dielectric is much faster than the internal dynamics of the electron-hole pair in the exciton, and the polarization rapidly follows the charged species. In this case, the effects of the polarization can be absorbed into a *static* screened electron-hole interaction. In contrast, because the binding energy of the lowest optically allowed exciton of an organic semiconductor is comparable to its excitation energy the effects of the polarization on this state must be modelled by a *dynamic* screened electron-hole interaction.

Moore and Yaron (1998) modelled screening processes in polyacetylene by surrounding a central solute chain by an increasing numbers of solvent chains arranged in a crystalline structure and extrapolating to an infinite system. The central chain was treated with a Pariser-Parr-Pople model with double and single bond transfer integrals of $t_d = 2.581$ eV and $t_s = 2.228$ eV, respectively, and $U = 11.13$ eV. The solvent chains were treated in an independent-electron approximation. Coulomb interactions were included between all chains. Replacing the central solute chain with a point charge led to an estimate of about 1 eV for the solvation energy of a point charge. In the assumption of an infinitely fast dielectric, where the solvent polarization is equilibrated to the instantaneous position of the electron and hole, the solvation energy of a well separated electron and hole is about 1.9 eV. Using a model that approximates the dynamic response of the dielectric, the free electron and hole become dressed by the dielectric response of the solvent chains to form polarons, and the solvation energy drops to about 1.5 eV. They also find in the dynamical model that the 1^1B_u state is screened by ~ 0.3 eV and the m^1Ag state is screened by ~ 0.6 eV. This means that a single chain calculation of the 1^1B_u binding energy is ~ 1.2 eV larger than its value for a polymer chain in the solid state.

Similar energetic corrections were found in a one-dimensional dynamical model of the dielectric (Barford *et al.* 2004). There it was also shown that covalent states are weakly screened by the environment, in contrast to the ionic states that are strongly screened.

9.5   Electron transfer

Both intra and inter molecular electron transfer are fundamental processes in electron transport in polymers. This in turn determines the operation of organic optoelectronic devices. Electron transfer between molecular complexes is also an important process in many biological systems. In this section we give a brief review of the nonadiabatic electron transfer that is relevant to electron transport. Since this process is nonadiabatic, the configuration coordinates of the initial and final electronic states are different. In practice, they are often taken to have the same values; namely the position at which the potential energy surfaces cross (called the reaction coordinates), as illustrated in Fig. 9.6.

In this section we derive expressions for the rates of electron transfer within the Fermi Golden Rule approximation. As we described for exciton transport in Section 9.2.4, these rates can be used to model charge transport using the density matrix formalism. There is a wide and thorough discussion of this topic in May and Kühn (2000).

9.5.1   *Unimolecular electron transfer*

We first consider electron transfer within a charged system composed of two strongly coupled subsystems: a donor, denoted by D, and an acceptor, denoted by A. The electron transfer occurs from D to A. These subsystems might be two parts of a molecular complex, or two moities of the same molecule, for example. Since these subsystems are strongly coupled they share the same nuclear normal coordinates. The adiabatic potential energy of the system is illustrated in Fig. 9.6. $U_D(Q)$ is the potential energy of the system when the charge is localized on D, while $U_A(Q)$ is the potential energy of the system when the charge is localized on A.

9.5.1.1   *Low temperature limit*   ($k_B T \ll \hbar\omega_\xi$ for all normal modes $\{\xi\}$) In this limit electron transfer occurs via nuclear tunnelling between the minima of U_D and U_A.

The initial, $|D\rangle$, and final, $|A\rangle$, states represent the electron localized on D and A, respectively. Within the Born-Oppenheimer approximation the electronic and nuclear degrees of freedom are described by the Born-Oppenheimer states,

$$|X\rangle = |x; \{Q\}\rangle |\nu_x\rangle. \tag{9.70}$$

As usual, $|x; \{Q\}\rangle$ is the electronic state, $|\nu_x\rangle$ is its associated nuclear state and the $\{Q\}$ label indicates that the electronic state is parametrized by the nuclear coordinates. Thus,

$$|D\rangle = |d; Q\rangle |\nu_d\rangle \tag{9.71}$$

and

$$|A\rangle = |a; Q\rangle |\nu_a\rangle, \tag{9.72}$$

for the simplified case of a single normal coordinate. The electron transfer integral, T_{DA}, is

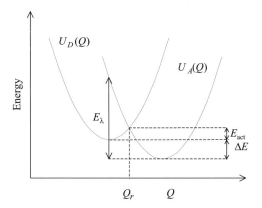

FIG. 9.6. The adiabatic potential energies of the donor-acceptor system. $U_D(Q)$ is the potential energy of the system when the charge is localized on the donor and $U_A(Q)$ is the potential energy of the system when the charge is localized on acceptor. E_{act}, E_λ and ΔE are the activation, reorganization and driving energies, as described in the text. The reaction coordinate, Q_r, is illustrated.

$$T_{DA} = \langle A|H^{(1)}|D\rangle, \qquad (9.73)$$

where $H^{(1)}$ is the one-electron Hamiltonian,

$$H^{(1)} = - \sum_{i\in D j \in A,\sigma} t_{ij} \left(c^\dagger_{i\sigma} c_{j\sigma} + c^\dagger_{j\sigma} c_{i\sigma} \right). \qquad (9.74)$$

Inserting eqns (9.71) and (9.72) into eqn (9.73) gives,

$$T_{DA} = \langle a; Q|H^{(1)}|d; Q\rangle \langle \nu_a|\nu_d\rangle, \qquad (9.75)$$

where we define the electronic matrix element as,

$$t_{DA} = \langle a; Q|H^{(1)}|d; Q\rangle. \qquad (9.76)$$

The nonadiabatic electron transfer rate is then given by the Fermi Golden Rule expression,

$$\begin{aligned} k_{DA} &= \frac{2\pi}{\hbar} \sum_{\nu_d \nu_a} f(E_{\nu_d})|T_{DA}|^2 \delta(E_{\nu_d} - E_{\nu_a}) \\ &= \frac{2\pi|t_{DA}|^2}{\hbar} \sum_{\nu_d \nu_a} f(E_{\nu_d}) F_{ad} \delta(E_{\nu_d} - E_{\nu_a}). \end{aligned} \qquad (9.77)$$

$f(E_{\nu_d})$ is the Bose distribution function for the occupancy of the ν_dth vibrational level and $F_{ad} = \langle \nu_a|\nu_d\rangle^2$ is the Franck-Condon factor. The sum is over all the initially populated levels ν_d and all the final levels, ν_a, satisfying energy conservation.

9.5.1.2 *High temperature limit* ($k_BT \gg \hbar\omega_\xi$ for all normal modes $\{\xi\}$) Formally, this limit may be obtained directly from the low-temperature expression. However, in this section we derive it classically. In the high-temperature limit electron transfer is an activated process due to thermal excitation over the potential barrier separating the two minima of the adiabatic potentials U_D and U_A, as illustrated in Fig. 9.6.

The nuclear coordinates are treated classically, via the configuration coordinate, Q. Then, the transfer rate is given by the Fermi Golden Rule expression,

$$k_{DA} = \frac{2\pi|t_{DA}|^2}{\hbar} \int f(Q)\delta(U_D(Q) - U_A(Q))dQ. \qquad (9.78)$$

$f(Q)$ is the *classical* (Boltzmann) distribution function,

$$f(Q) = \sqrt{\frac{M\omega^2}{2\pi k_BT}} \exp\left(-\frac{U_D(Q)}{k_BT}\right), \qquad (9.79)$$

which ensures an ensemble average over the initial configurations of the donor.

Evaluating the integral in eqn (9.78) gives,

$$k_{DA} = |t_{DA}|^2 \sqrt{\frac{\pi}{\hbar^2 k_BTE_\lambda}} \exp\left(-\frac{E_{act}}{k_BT}\right), \qquad (9.80)$$

where E_{act} is the *activation energy* and E_λ is the *reorganization energy*, both of which are defined in Fig. 9.6. E_λ is the energy lost when there is a vertical transition from D to A, followed by vibrational relaxation to the minimum of $U_A(Q)$.[39]

It is more convenient to re-express the activation energy in terms of the reorganization energy and *driving energy*, ΔE, (also defined in Fig. 9.6), as follows.

$$E_{act} = \frac{(\Delta E - E_\lambda)^2}{4E_\lambda}. \qquad (9.81)$$

Then eqn (9.80) becomes the Markus expression (Markus 1964) for electron transfer,

$$k_{DA} = |t_{DA}|^2 \sqrt{\frac{\pi}{\hbar^2 k_BTE_\lambda}} \exp\left(-\frac{(\Delta E - E_\lambda)^2}{4k_BTE_\lambda}\right). \qquad (9.82)$$

It is customary to consider electron transfer in three limits. (a) The *normal limit*, defined by $E_\lambda > \Delta E$, (b) the *activationless limit*, defined by $E_\lambda = \Delta E$, and (c) the *inverted limit*, defined by $E_\lambda < \Delta E$. These limits are illustrated in Fig. 9.7.

---

[39] As expected from detailed balance, $k_{AD} = k_{DA}\exp(-\Delta E/k_BT)$.

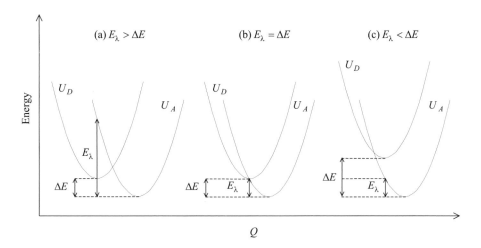

FIG. 9.7. The adiabatic potential energies in the (a) normal limit, (b) activationless limit and (c) inverted limit.

9.5.2 *Bimolecular electron transfer*

Bimolecular electron transfer occurs between two separate, weakly-coupled molecules with independent nuclear degrees of freedom. This is the more appropriate description for charge transfer between polymers in solution and weakly coupled polymers in a thin film. Electron transfer will occur from a charged donor molecule to a neutral acceptor molecule. The adiabatic potential energies of the molecules in their charge and neutral states are illustrated in Fig. 9.8.

9.5.2.1 *Low temperature limit* We consider electron transfer from a donor molecule initially in a negative charge state, $|D^-\rangle$, to an acceptor molecule initially in a neutral state, $|A\rangle$. The initial and final states are therefore

$$|I\rangle = |D^-\rangle|A\rangle \tag{9.83}$$

and

$$|F\rangle = |D\rangle|A^-\rangle, \tag{9.84}$$

respectively, where,

$$|D^-\rangle = |d^-; Q_D\rangle|\nu_{d-}\rangle, \tag{9.85}$$

$$|D\rangle = |d; Q_D\rangle|\nu_d\rangle, \tag{9.86}$$

$$|A\rangle = |a; Q_A\rangle|\nu_a\rangle, \tag{9.87}$$

and

$$|A^-\rangle = |a^-; Q_A\rangle|\nu_{a-}\rangle. \tag{9.88}$$

Q_D and Q_A are the nuclear configuration coordinates for the donor and acceptor, respectively.

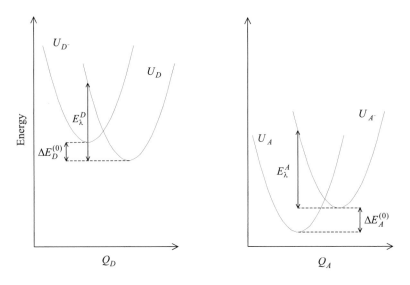

FIG. 9.8. The adiabatic potential energies of the donor (D) and acceptor (A) molecules in their charged and neutral states.

Electron transfer between molecules is mediated by $H_\perp$ defined by,

$$H_\perp = -t_{ij} \sum_{i\in Dj\in A,\sigma} \left(c_{i\sigma}^{D\dagger} c_{j\sigma}^{A} + c_{j\sigma}^{A\dagger} c_{i\sigma}^{D} \right), \tag{9.89}$$

where $c_{i\sigma}^{X\dagger}$ ($c_{i\sigma}^{X}$) creates (destroys) a π-electron on site i of molecule X and t_{ij} is the interchain hybridization integral. Hence, the electron transfer integral is

$$T_{DA} = \langle F|H_\perp|I\rangle = t_{DA}\langle \nu_{a-}|\nu_a\rangle\langle \nu_d|\nu_{d-}\rangle, \tag{9.90}$$

where the electronic matrix element is

$$t_{DA} = \sum_{i\in Aj\in B,\sigma} t_{ij}\langle a^-; Q_A|c_{j\sigma}^{A\dagger}|a; Q_A\rangle\langle d; Q_D|c_{i\sigma}^{D}|d^-; Q_D\rangle. \tag{9.91}$$

The electron transfer rate is now,

$$k_{DA} =$$
$$\frac{2\pi|t_{DA}|^2}{\hbar} \sum_{\nu_{d-}\nu_d} \sum_{\nu_a\nu_{a-}} f(E_{\nu_{d-}})f(E_{\nu_a})F_{dd-}F_{a-a}\delta((E_{\nu_{d-}}+E_{\nu_a})-(E_{\nu_d}+E_{\nu_{a-}})), \tag{9.92}$$

where $f(E_{\nu_{d-}})$ and $f(E_{\nu_a})$ are the Bose distribution functions for the initial population of the vibrational levels, while $F_{dd-} = \langle \nu_d|\nu_{d-}\rangle^2$ and $F_{a-a} = \langle \nu_{a-}|\nu_a\rangle^2$. Using the identity in eqn (9.34) we can rewrite the δ-function as,

$$\delta((E_{\nu_{d-}} + E_{\nu_a}) - (E_{\nu_d} + E_{\nu_{a-}})) = \int \delta(E_{\nu_{d-}} - E_{\nu_d} - \hbar\omega)\delta(E_{\nu_a} - E_{\nu_{a-}} + \hbar\omega)d(\hbar\omega).$$

(9.93)

Then, defining the spectral functions $D(\omega)$ and $A(\omega)$ as

$$D(\omega) = \sum_{\nu_{d-}\nu_d} f(E_{\nu_{d-}})F_{dd-}\delta(E_{\nu_{d-}} - E_{\nu_d} - \hbar\omega)$$

(9.94)

and

$$A(\omega) = \sum_{\nu_a\nu_{a-}} f(E_{\nu_a})F_{a-a}\delta(E_{\nu_a} - E_{\nu_{a-}} + \hbar\omega)$$

(9.95)

the electron transfer rate may be expressed as,

$$k_{DA} = 2\pi|t_{DA}|^2 \int D(\omega)A(\omega)d\omega.$$

(9.96)

As in exciton transfer, the overlap of the spectral functions reflects energy conservation during the electron transfer.

9.5.2.2 *High temperature limit* The electron transfer rate in the high temperature limit follows in a similar manner to that of Section 9.5.1.2. Now, however, there are two independent configuration coordinates, Q_D and Q_A.

The transfer rate is thus,

$$k_{DA} = \frac{2\pi|t_{DA}|^2}{\hbar} \int f(Q_D)f(Q_A)\delta((U_{D-}(Q) + U_A(Q)) - (U_D(Q) + U_{A-}(Q)))dQ_D dQ_A.$$

(9.97)

Again, using the identity eqn (9.34) we define

$$D(\omega) = \int f(Q_D)\delta(U_{D-}(Q) - U_D(Q) - \hbar\omega)dQ_D$$

$$\equiv \sqrt{\frac{\pi}{\hbar^2 k_B T E_\lambda^D}} \exp\left(-\frac{(\Delta E_D - E_\lambda^D)^2}{4k_B T E_\lambda^D}\right)$$

(9.98)

and

$$A(\omega) = \int f(Q_A)\delta(U_A(Q) - U_{A-}(Q) + \hbar\omega)dQ_A$$

$$\equiv \sqrt{\frac{\pi}{\hbar^2 k_B T E_\lambda^A}} \exp\left(-\frac{(\Delta E_A - E_\lambda^A)^2}{4k_B T E_\lambda^A}\right),$$

(9.99)

where

$$\Delta E^D = \Delta E_D^{(0)} - \hbar\omega,$$

(9.100)

$$\Delta E^A = \Delta E_A^{(0)} + \hbar\omega, \tag{9.101}$$

and E_λ^D, E_λ^A, $\Delta E_D^{(0)}$, and $\Delta E_A^{(0)}$ are defined in Fig. 9.8. Then we obtain an identical expression for the electron transfer rate as in the last section, namely,

$$k_{DA} = 2\pi |t_{DA}|^2 \int D(\omega) A(\omega) d\omega. \tag{9.102}$$

Bimolecular electron transfer is an important mechanism in determining the singlet exciton yield in light emitting polymers, as discussed in the next section.

9.6   The singlet exciton yield in light emitting polymers

9.6.1   *Introduction*

In the final section of this Chapter we discuss the singlet exciton yield in light emitting polymers obtained via charge injection. As we now show, this yield is a key a factor in the electroluminescence efficiency of polymer light emitting devices.

The operation of a light emitting device is shown schematically in Fig. 9.9. Electrons and holes are injected into the conduction and valence bands at the cathode and anode, respectively. Under the influence of the electric field they drift through the device, rapidly being captured and (as explained below) eventually forming the lowest energy singlet or triplet excitons by interconversion. The singlet excitons predominately recombine radiatively, whereas the triplet excitons recombine nonradiatively.

The electroluminescence quantum efficiency, η_{EL}, of a light emitting device is defined as,

$$\eta_{EL} = \left(\frac{\text{Rate of photons detected}}{I}\right), \tag{9.103}$$

where I is the current of injected electron-hole pairs. It is convenient to factorize this expression into three factors,

$$\eta_{EL} = \left(\frac{\text{Number of detected photons}}{\text{Number of emitted photons}}\right)$$
$$\times \left(\frac{\text{Number of emitted photons}}{\text{Number of singlet excitons}}\right)\left(\frac{\text{Rate of singlet exciton decay}}{I}\right). \tag{9.104}$$

Finally, we define the singlet exciton yield, η_S, by,

$$\eta_S = \frac{1}{\eta_c}\left(\frac{\text{Rate of singlet exciton decay}}{I}\right), \tag{9.105}$$

where η_c is the fraction of electron-hole pairs captured.

Since electrons and holes are ejected into the device with random spin orientations, there are initially three times as many triplet electron-hole pairs as

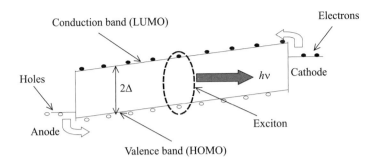

FIG. 9.9. The schematic operation of a light emitting polymer device. Electrons in the conduction band and holes in the valence band are mutually captured, forming strongly bound exciton states. However, only the singlet excitons decay radiatively, emitting a photon. Triplet excitons decay nonradiatively, emitting phonons that is eventually wasted as heat.

singlet electron-hole pairs. Thus, when there is spin-independent recombination, $\eta_S = 25\%$. Singlet exciton yields in light emitting polymer devices exceeding the spin-independent recombination value of 25% have now been reported by a large number of groups (Cao *et al.* 1999; Ho *et al.* 2000; Dhoot *et al.* 2002; Wilson *et al.* 2001; Wohlgenannt *et al.* 2002). Although the value of η_S remains controversial (see Segal *et al.* 2003), an understanding and control of it is an important factor in optimizing device performance.

Various theoretical attempts have been made to explain the enhanced singlet exciton yield, including models based on either intramolecular and intermolecular recombination. Bittner *et al.* (see Karabunarliev and Bittner (2003)) assume that intrachain electron-hole recombination occurs via vibrational relaxation through the band of exciton states between the particle-hole continuum and lowest bound excitons. Since vibrational relaxation is faster in the singlet channel than the triplet channel, because the lowest singlet exciton lies higher in energy than the lowest triplet exciton, a faster formation rate for the singlet than the triplet exciton is predicted. Hong and Meng (2001) argue that a multiphonon process in the triplet channel also leads to faster intramolecular singlet exciton formation.

The different rates for singlet and triplet exciton formation predicted in the literature for interchain recombination (Ye *et al.* 2002; Tandon *et al.* 2003) arise largely from the assumption that an interchain density-dependent electron transfer term is an important factor in the recombination mechanism. This term couples states of the same ionicity. Since the interchain charge transfer states are predominately ionic, while the intrachain triplet exciton has more covalent character than the intrachain singlet exciton, the rate for the singlet exciton formation is correspondingly greater.

The recent experimental and theoretical work is reviewed in Wohlgenannt and Vardeny (2003), and Köhler and Wilson (2003).

In this section we describe a model of interchain electron-hole recombination that involves intermediate, loosely bound (charge-transfer) states that lie energetically between the electron-hole continuum and the final, strongly bound exciton states (Barford 2004). We will show that intermolecular interconversion from the charge transfer to the lowest energy exciton states is limited by multiphonon emission, which decreases approximately exponentially with the energy gap between the pair of states. Since the lowest singlet and triplet exciton energies are split by a large exchange energy (of ca. 0.7 eV), while the charge-transfer states are quasi-degenerate, the triplet exciton formation rate is considerably smaller than the singlet exciton rate. Thus the singlet exciton yield may be greater than 25%, provided that the intersystem crossing mechanisms are fast enough.

The theory of this section relies on the assumption that excitons in light emitting polymers are Mott-Wannier excitons (as described in Chapter 6). The experimental and theoretical evidence for this assumption is described in detail in Chapter 11.

9.6.2 *Basic model and the rate equations*

Figure 9.10 shows the energy level diagram for this model. The electrons and holes are injected into the polymer device with random spin orientations. Under the influence of the electric field the electrons and holes migrate through the device, rapidly being captured (in less than 10^{-12} s) to form the weakly bound charge-transfer singlet and triplet excitons, S_{CT} and T_{CT}, respectively. We assume that no spin mixing occurs during this process, and thus the ratio of S_{CT} to T_{CT} is 1:3.[40] If the intersystem crossing (ISC) between T_{CT} and S_{CT} (with a rate $1/\tau_{\mathrm{ISC}}$) competes with the interconversion from T_{CT} to the triplet exciton, T_{X}, (with a rate $1/\tau_{T_{\mathrm{CT}}}$) and $1/\tau_{T_{\mathrm{CT}}}$ is smaller than the interconversion rate ($1/\tau_{S_{\mathrm{CT}}}$) from S_{CT} to the singlet exciton, S_{X}, then the singlet yield is enhanced.

The charge-transfer states might be either intramolecular loosely bound excitons or weakly bound positive and negative polarons on neighbouring chains. Intramolecular charge-transfer states are Mott-Wannier excitons whose relative electron-hole wavefunctions are odd under electron-hole exchange. As explained in Chapter 6, a crucial characteristic of these states is that, because of their odd electron-hole parity, the probability of finding the electron and hole on the same molecular repeat-unit is zero. Thus, they experience very small exchange interactions and therefore the singlet and triplet states are quasi-degenerate.[41] Similarly, the intermolecular weakly bound positive and negative polarons - although now possessing even electron-hole parity - also experience weak exchange interactions as necessarily the electron and hole are on different repeat units, and thus singlet and triplet states are also quasi-degenerate. Furthermore, since the charge-transfer excitons are also weakly bound with relatively large electron-hole

---

[40]Furthermore, there is an efficient electric-field-induced electron-hole mixing, so all the electron-hole pairs become charge-transfer excitons.

[41]This prediction is confirmed in PPV-DOO, where $E(S_{\mathrm{CT}}) \sim 3.2$ eV (Frolov *et al.* 2002) and $E(T_{\mathrm{CT}}) \sim 3.1$ eV (Monkman *et al.* 2001). See also Kadashchuk *et al.* (2004).

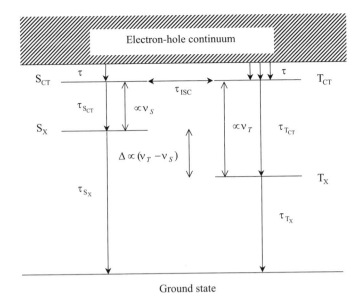

FIG. 9.10. The energy level diagram of the quasi-degenerate singlet and triplet charge-transfer excitons (denoted by S_{CT} and T_{CT}, respectively) and the lowest singlet and triplet excitons (denoted by S_X and T_X, respectively). S_{CT} and T_{CT} may be either intramolecular odd parity excitons or intermolecular even parity excitons. (In each case these correspond to the lowest pseudo-momentum members of each exciton family, as described in Section 9.6.3.1.) Also shown are the respective lifetimes (or inverse rates) for the interconversions within the same spin manifolds and intersystem crossing (ISC) between the spin manifolds. Δ is the exchange energy between S_X and T_X.

separations, there exist efficient spin-flipping mechanisms, such as spin-orbit coupling, or exciton disassociation via the electric field or by scattering from free carriers and defects. Here we focuss on interconversion to the strongly bound excitons from the *interchain* charge-transfer excitons.

The strongly bound excitons, S_X and T_X, are intramolecular states. The interconversion process from S_{CT} to S_X and from T_{CT} to T_X depends on the nature of S_{CT} and T_{CT}. The mechanism for bimolecular interconversion is described more fully in the next section. In this section we describe the kinetics by classical rate equations. The use of classical rate equations is justified if rapid interconversion follows the ISC between T_{CT} and S_{CT}, as then there will be no coherence or recurrence between T_{CT} and S_{CT}. We also note that since interconversion is followed by rapid vibrational-relaxation (in a time of $\sim 10^{-13}$ s) these processes are irreversible.

We first consider the case where ISC occurs directly via the spin-orbit coupling operator. This operator converts the $S_z = \pm 1$ triplets into the singlet, and vice versa. Let N_{S_X}, $N_{S_{CT}}$, $N_{T_X}^{\pm}$, and $N_{T_{CT}}^{\pm}$ denote the number of S_X, S_{CT}, and the $S_z = \pm 1$ T_X and T_{CT} excitons, respectively. $N/\tau = I\eta_c$ is the number electron-hole pairs created per second (where η_c is the fraction of electron-hole pairs captured and I is the current of electron-hole pairs). Then the rate equations are:

$$\frac{dN_{S_{CT}}}{dt} = \frac{N}{4\tau} + \frac{N_{T_{CT}}^{\pm}}{\tau_{ISC}} - N_{S_{CT}}\left(\frac{1}{\tau_{ISC}} + \frac{1}{\tau_{S_{CT}}}\right), \tag{9.106}$$

$$\frac{dN_{T_{CT}}^{\pm}}{dt} = \frac{N}{2\tau} + \frac{N_{S_{CT}}}{\tau_{ISC}} - N_{T_{CT}}^{\pm}\left(\frac{1}{\tau_{ISC}} + \frac{1}{\tau_{T_{CT}}}\right), \tag{9.107}$$

$$\frac{dN_{S_X}}{dt} = \frac{N_{S_{CT}}}{\tau_{S_{CT}}} - \frac{N_{S_X}}{\tau_{S_X}}, \tag{9.108}$$

and

$$\frac{dN_{T_X}^{\pm}}{dt} = \frac{N_{T_{CT}}^{\pm}}{\tau_{T_{CT}}} - \frac{N_{T_X}^{\pm}}{\tau_{T_X}}. \tag{9.109}$$

Notice that the $S_z = 0$ component of the T_{CT} exciton is *converted directly* to the $S_z = 0$ component of the T_X exciton, and cannot contribute to the singlet exciton yield.

From eqn (9.105) the singlet exciton yield, η_S, is defined by,

$$\eta_S = \frac{N_{S_X}/\tau_{S_X}}{I\eta_c} \equiv \frac{N_{S_X}/\tau_{S_X}}{N/\tau}. \tag{9.110}$$

Solving the rate equations under the steady state conditions that

$$\frac{dN_{S_{CT}}}{dt} = \frac{dN_{T_{CT}}^{\pm}}{dt} = \frac{dN_{S_X}}{dt} = \frac{dN_{T_X}^{\pm}}{dt} = 0 \tag{9.111}$$

gives

$$\eta_S = \frac{3+\gamma}{4(1+\beta+\gamma)}, \tag{9.112}$$

where $\beta = \tau_{S_{CT}}/\tau_{T_{CT}}$ and $\gamma = \tau_{ISC}/\tau_{T_{CT}}$.

Alternatively, we might consider ISC via a spin-randomization process, whereby the charge-transfer excitons are scattered into charge-transfer triplets with a probability of 3/4 and charge-transfer singlets with a probability of 1/4. Then the rate equations are (Wohlgenannt and Vardeny 2003),

$$\frac{dN_{S_{CT}}}{dt} = \frac{N}{4}\left(\frac{1}{\tau} + \frac{1}{\tau_{ISC}}\right) - N_{S_{CT}}\left(\frac{1}{\tau_{ISC}} + \frac{1}{\tau_{S_{CT}}}\right), \tag{9.113}$$

$$\frac{dN_{T_{CT}}}{dt} = \frac{3N}{4}\left(\frac{1}{\tau} + \frac{1}{\tau_{ISC}}\right) - N_{T_{CT}}\left(\frac{1}{\tau_{ISC}} + \frac{1}{\tau_{T_{CT}}}\right), \tag{9.114}$$

$$\frac{dN_{S_X}}{dt} = \frac{N_{S_{CT}}}{\tau_{S_{CT}}} - \frac{N_{S_X}}{\tau_{S_X}}, \tag{9.115}$$

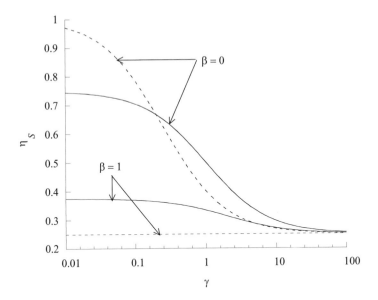

FIG. 9.11. The singlet exciton yield, η_S, versus $\gamma = \tau_{\mathrm{ISC}}/\tau_{T_{\mathrm{CT}}}$. $\beta = \tau_{S_{\mathrm{CT}}}/\tau_{T_{\mathrm{CT}}}$. Solid curves from eqn (9.112), dashed curves from eqn (9.117).

and

$$\frac{dN_{T_{\mathrm{X}}}}{dt} = \frac{N_{T_{\mathrm{CT}}}}{\tau_{T_{\mathrm{CT}}}} - \frac{N_{T_{\mathrm{X}}}}{\tau_{T_{\mathrm{X}}}}. \tag{9.116}$$

where $N_{T_{\mathrm{CT}}}$ is the total number of charge-transfer triplets. Solving again for the steady state, the singlet exciton yield becomes,

$$\eta_S = \frac{1 + \gamma}{1 + 3\beta + 4\gamma}, \tag{9.117}$$

In practice, as we shall show, $\tau_{T_{\mathrm{CT}}} \gg \tau_{S_{\mathrm{CT}}}$ so $\beta \approx 0$. We note that η_S is a function only of the relative life-times of charge-transfer singlets and triplets, and the ISC rate. The singlet yield is plotted in Fig. 9.11 as a function of γ. We now describe the calculation of the relative rates.

9.6.3 Derivation of the intermolecular interconversion rate

Intermolecular interconversion occurs via the electron-transfer Hamiltonian, $H_\perp$. For parallel chains with nearest neighbour electron transfer this is

$$H_\perp = -t_\perp \sum_{i\sigma} \left(c_{i\sigma}^{(1)\dagger} c_{i\sigma}^{(2)} + c_{i\sigma}^{(2)\dagger} c_{i\sigma}^{(1)} \right), \tag{9.118}$$

where $c_{i\sigma}^{(\ell)\dagger}$ ($c_{i\sigma}^{(\ell)}$) creates (destroys) a π-electron on site i of chain ℓ and $t_\perp$ is the interchain hybridization integral.

If the chains are weakly coupled we may regard $H_\perp$ as a perturbation on the approximate Hamiltonian, H_0, where H_0 contains the intramolecular Hamiltonians and the remaining interchain interactions - particularly the Coulomb interactions. Thus, we may write the Born-Oppenheimer Hamiltonian for a pair of coupled polymer chains as,

$$H = H_0 + H_\perp. \tag{9.119}$$

Within the Born-Oppenheimer approximation the electronic and nuclear degrees of freedom are described by the Born-Oppenheimer states,

$$|A\rangle = |a; \{Q\}\rangle|\nu_a\rangle, \tag{9.120}$$

where $|a; \{Q\}\rangle$ is the electronic state and $|\nu_a\rangle$ is its associated nuclear state.

We assume that the stationary electronic states are the eigenstates of the approximate Hamiltonian, H_0. Thus, the perturbation, $H_\perp$, mixes these electronic states. In particular, it causes an interconversion from the interchain excitons (or weakly bound polaron pairs) to the intrachain excitons.

We take the initial electronic state, $|i\rangle$, to be a positive polaron on chain 1 and a negative polaron on chain 2,

$$|i\rangle = |P^+, P^-; Q_1, Q_2\rangle. \tag{9.121}$$

The interchain Coulomb interaction between the chains creates a weakly bound charge-transfer exciton, to be described below. The labels Q_1 and Q_2 indicate the independent normal coordinates of chains 1 and 2, respectively.

The effect of $H_\perp$ is to move charge from one chain to another. We consider the situation where the negative polaron is transferred from chain 2 to chain 1. Thus, the final electronic state, $|f\rangle$, is an intramolecular exciton on chain 1 (denoted by $|EX\rangle$), leaving chain 2 in its ground electronic state,

$$|f\rangle = |EX; Q_1\rangle^{(1)}|GS; Q_2\rangle^{(2)}. \tag{9.122}$$

These electronic states are illustrated in Fig. 9.12.

The isoenergetic interconversion rate from the initial to the final states is determined by the Fermi Golden Rule expression,

$$k_{I \to F} = \frac{2\pi}{\hbar} \langle F|H_\perp|I\rangle^2 \delta(E_F - E_I), \tag{9.123}$$

where the initial and final Born-Oppenheimer states are,

$$|I\rangle = |P^+, P^-\rangle|\nu_{P+}\rangle^{(1)}|\nu_{P-}\rangle^{(2)} \tag{9.124}$$

and

$$|F\rangle = |EX\rangle^{(1)}|GS\rangle^{(2)}|\nu_{EX}\rangle^{(1)}|\nu_{GS}\rangle^{(2)}, \tag{9.125}$$

respectively. The matrix element appearing in eqn (9.123) is a product of the electronic matrix elements and the vibrational overlaps,

Initial state: $|i\rangle = |P^+, P^-\rangle = |\tilde{\psi}_{n=1}\Psi_{j=1}\rangle$

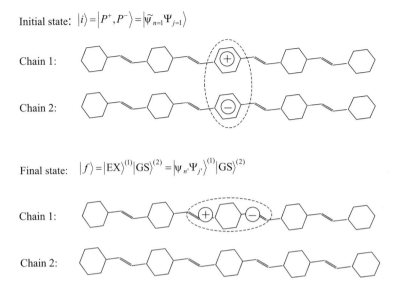

Chain 1:

Chain 2:

Final state: $|f\rangle = |EX\rangle^{(1)}|GS\rangle^{(2)} = |\psi_{n'}\Psi_{j'}\rangle^{(1)}|GS\rangle^{(2)}$

Chain 1:

Chain 2:

FIG. 9.12. A schematic representation of the initial and final electronic states. The initial state, $|i\rangle$, is an electron and hole on neighbouring chains in a weakly bound intermolecular charge transfer state. The final state, $|f\rangle$, is an electron and hole on chain 1 bound in an intramolecular exciton state, while chain 2 is in its electronic ground state.

$$\langle F|H_\perp|I\rangle = \langle f|H_\perp|i\rangle^{(1)}\langle\nu_{EX}|\nu_{P+}\rangle^{(1)(2)}\langle\nu_{GS}|\nu_{P-}\rangle^{(2)}. \qquad (9.126)$$

We derive expressions for the electronic matrix element and the vibrational overlaps in the following two sections.

9.6.3.1 *Electronic matrix elements* The corresponding electronic matrix element is

$$\langle f|H_\perp|i\rangle = {}^{(2)}\langle GS|^{(1)}\langle EX|H_\perp|P^+, P^-\rangle. \qquad (9.127)$$

We evaluate this matrix element using the effective-particle exciton model introduced in Chapter 6. We briefly review this theory here. In the weak-coupling limit[42] (namely, the limit that the Coulomb interactions are less than or equal to the band width) the intramolecular excited states of semiconducting conjugated polymers are Mott-Wannier excitons described by,

$$|EX\rangle = \sum_{r,R} \Phi_{nj}(r, R)|r, R\rangle. \qquad (9.128)$$

$|r, R\rangle$ is an electron-hole basis state (defined in eqn (6.7)) constructed by promoting an electron from the filled valence band Wannier orbital at $R - r/2$ to

[42]This is the relevant limit for light emitting polymers, as explained in Chapter 11.

the empty conduction band Wannier orbital at $R + r/2$. R is the center-of-mass coordinate and r is the relative coordinate of the effective particle.

$\Phi_{nj}(r, R)$ is the total exciton wavefunction. In the effective-particle approximation this may be expressed as a product of the relative wavefunction, $\psi_n(r)$, and the centre-of-mass wavefunction, $\Psi_j(R)$:

$$\Phi_{nj}(r, R) = \psi_n(r)\Psi_j(R). \tag{9.129}$$

$\psi_n(r)$ is a hydrogen-like electron-hole wavefunction labelled by the principle quantum number, n. This has the property that under electron-hole reflection (namely, $r \to -r$) $\psi_n(r) = \psi_n(-r)$ for odd n and $\psi_n(r) = -\psi_n(-r)$ for even n. On a linear chain the centre-of-mass wavefunction is

$$\Psi_j(R) = \sqrt{\frac{2}{N+1}} \sin(\beta_j R), \tag{9.130}$$

where N is the number of unit cells. For each principle quantum number, n, there is a band of excitons with different pseudo-momentum, $\beta_j = j\pi/(N+1)d$, where j satisfies $1 \le j \le N$ and d is the unit cell distance. Thus, every exciton state label, EX, corresponds to two independent quantum numbers: n and j.

As described in Chapter 6, $n = 1$ corresponds to the S_X and T_X families of intrachain excitons, while $n = 2$ corresponds to the S_{CT} and T_{CT} families of *intrachain* excitons. The lowest energy branch of each family has the smallest pseudo-momentum, namely, $j = 1$. The effective-particle model is illustrated in Fig. 6.2.

It is also convenient to describe the intermolecular weakly bound polaron pairs as charge-transfer excitons described by

$$|P^+, P^-\rangle = \sum_{r,R} \tilde{\psi}_n(r)\Psi_j(R)|r, R; 2\rangle, \tag{9.131}$$

where $\tilde{\psi}_n$ represents the interchain effective-particle wavefunction. $n = 1$ (namely, even electron-hole parity) for the lowest energy interchain excitons. $|r, R; 2\rangle$ is an electron-hole basis state constructed by promoting an electron from the filled valence band Wannier orbital at $R - r/2$ on chain 1 to the empty conduction band Wannier orbital at $R + r/2$ on chain 2.

Using eqns (9.128), (9.129), and (9.131) the electronic matrix element is

$$\langle f|H_\perp|i\rangle = \sum_{r',R'} \psi_{n'}(r')\Psi_{j'}(R') \sum_{r,R} \tilde{\psi}_n(r)\Psi_j(R) {}^{(2)}\langle GS|^{(1)}\langle r', R'|H_\perp|r, R; 2\rangle. \tag{9.132}$$

This matrix element is evaluated by expressing $H_\perp$ in terms of the valence and conduction Wannier orbital operators. Retaining terms that keep within the exciton subspace $H_\perp$ becomes

$$H_\perp = -t_\perp \sum_{j\sigma} \left(c^{v(1)\dagger}_{j\sigma} c^{v(2)}_{j\sigma} + c^{c(1)\dagger}_{j\sigma} c^{c(2)}_{j\sigma} \right) + \text{H.C.} \qquad (9.133)$$

Then,

$$\langle f|H_\perp|i\rangle = -t_\perp \sum_{r',r} \psi_{n'}(r')\tilde{\psi}_n(r) \sum_{R',R} \Psi_{j'}(R')\Psi_j(R)^{(1)}\langle r', R'|r, R\rangle^{(1)}. \qquad (9.134)$$

By exploiting the orthonormality of the basis functions,

$$\langle r', R'|r, R\rangle = \delta_{r'r}\delta_{R'R}, \qquad (9.135)$$

we have

$$\langle f|H_\perp|i\rangle = -t_\perp \sum_r \psi_{n'}(r)\tilde{\psi}_n(r) \sum_R \Psi_{j'}(R)\Psi_j(R). \qquad (9.136)$$

Now, the centre-of-mass wavefunctions also form an orthonormal set,[43]

$$\sum_R \Psi_{j'}(R)\Psi_j(R) = \delta_{j'j}. \qquad (9.137)$$

Thus, using eqn (9.137) in eqn (9.136) we obtain the final result for the electronic matrix element,

$$\langle f|H_\perp|i\rangle = -t_\perp \sum_r \psi_{n'}(r)\tilde{\psi}_n(r). \qquad (9.138)$$

Equations (9.137) and (9.138) demonstrate the important result that inter-conversion via $H_\perp$ is subject to two electronic selection rules.

1. Interconversion occurs between excitons with the same centre-of-mass pseudo-momentum, β_j.
2. Interconversion occurs between excitons with the same electron-hole parity. Thus, $|n' - n| = $ even.

Since the lowest energy interchain excitons have even electron-hole parity this implies that $H_\perp$ connects them to S_X and T_X, and not to the intramolecular S_{CT} and T_{CT}.[44] Moreover, since the interchain exciton will have relaxed to its lowest momentum state, $H_\perp$ converts it to the intrachain exciton in its lowest momentum state, and *not to higher lying momentum states*. Interconversion subject to these selection rules is illustrated in Fig. 9.13.

[43] For example, the $\{\sin(\beta_j R)\}$ functions satisfy $2/(N+1) \sum_R \sin(\beta_{j'} R) \sin(\beta_j R) = \delta_{j'j}$.

[44] However, $H_\perp$ can connect the interchain excitons to other even parity intrachain excitons, if the former lie higher in energy. In that case the electron-hole recombination can be regarded as an intrachain processes, as the states must relax via the intrachain charge-transfer excitons, S_{CT} and T_{CT}.

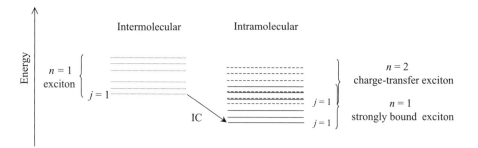

FIG. 9.13. Electronic selection rules determine that intermolecular interconversion occurs from the $n = 1$, $j = 1$ intermolecular exciton to the $n = 1$, $j = 1$ intramolecular exciton.

9.6.3.2 *Vibrational overlaps* We now discuss the contribution of the vibrational wavefunctions to the total matrix element, eqn (9.126).

Intermolecular interconversion is an isoenergetic process which occurs from the lowest pseudomomentum state of the charge-transfer manifold and the lowest vibrational levels of this state to the lowest pseudomomentum state of the intramolecular excitons at the same energy as the initial level. Thus, the vibrational levels in eqn (9.124) are $\nu_{P+} = 0$ and $\nu_{P-} = 0$. However, the vibrational levels in eqn (9.125) are determined by the conservation of energy.

Using eqn (9.123) the rate is thus,

$$
\begin{aligned}
k_{I \to F} &= \frac{2\pi}{\hbar} |\langle f|H_\perp|i\rangle|^2 \sum_{\nu_{EX}\nu_{GS}} F^{(1)}_{0\nu_{EX}} F^{(2)}_{0\nu_{GS}} \delta(E_F - E_I) \\
&= \frac{2\pi}{\hbar} |\langle f|H_\perp|i\rangle|^2 \sum_{\nu_{EX}\nu_{GS}} F^{(1)}_{0\nu_{EX}} F^{(2)}_{0\nu_{GS}} \delta(\Delta E_1^{\nu_{EX}} + \Delta E_2^{\nu_{GS}})
\end{aligned}
$$

$$(9.139)$$

where

$$
F^{(1)}_{0\nu_{EX}} = |{}^{(1)}\langle 0_{P+}|\nu_{EX}\rangle^{(1)}|^2 \tag{9.140}
$$

and

$$
F^{(2)}_{0\nu_{GS}} = |{}^{(2)}\langle 0_{P-}|\nu_{GS}\rangle^{(2)}|^2 \tag{9.141}
$$

are the Franck-Condon factors associated with the vibrational wavefunction overlaps of chains 1 and 2, respectively. Likewise,

$$
\Delta E_1^{\nu_{EX}} = E_1(EX; \nu_{EX}) - E_1(P^+; 0_{P+}) \tag{9.142}
$$

and

$$
\Delta E_2^{\nu_{GS}} = E_2(GS; \nu_{GS}) - E_2(P^-; 0_{P-}) \tag{9.143}
$$

are the changes in energy of chains 1 and 2, respectively. These changes in energy are illustrated in Fig. 9.14.

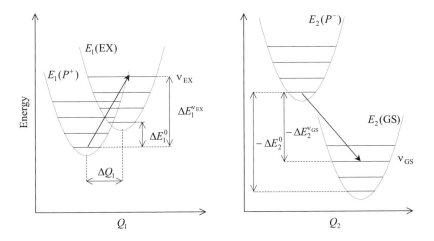

FIG. 9.14. Bimolecular electron transfer from chain 2 (right) to chain 1 (left). The initial state is a positive polaron on chain 1 and a negative polaron on chain 2, each in their lowest vibrational level. The electron transfer creates an exciton state in chain 1, with chain 2 in its ground state. The energy differences between the final and initial states are ΔE_ℓ^ν for the ℓth chain. From energy conservation, $\Delta E_2^{\nu_{GS}} = -\Delta E_1^{\nu_{EX}}$. If $\Delta Q_1 = 0$ then $\Delta E_1^{\nu_{EX}} = \Delta E_1^0$ and thus $\Delta E_2^{\nu_{GS}} = -\Delta E_1^0$.

Using the same procedure as Section 9.5.2 we can recast eqn (9.139) into the familiar rate expression for bimolecular electron transfer (see eqn (9.96)),

$$k_{I \to F} = \frac{2\pi}{\hbar} |\langle f|H_{\text{inter}}^1|i\rangle|^2 \int D(E)A(E)dE, \qquad (9.144)$$

where the spectral functions for the donor (chain 2) and acceptor (chain 1) are

$$D(E) = \sum_{\nu_{GS}} F_{0\nu_{GS}}^{(2)} \delta(\Delta E_2^{\nu_{GS}} + E) \qquad (9.145)$$

and

$$A(E) = \sum_{\nu_{EX}} F_{0\nu_{EX}}^{(1)} \delta(\Delta E_1^{\nu_{EX}} - E), \qquad (9.146)$$

respectively (where the expressions are taken at T = 0 K).

A useful simplification to this expression arises by noting that the geometric distortions of the polarons and exciton polarons (namely the 1^1B_u or 1^3B_u states) from the ground state structure are similar (as described in Chapter 7). Thus, the Huang-Rhys parameter (proportional to ΔQ_1^2, as defined in Fig. 9.14) for the 1^1B_u and 1^3B_u states relative to the positive polaron is negligible.[45] Therefore, to a good approximation,

[45]This assumption is less valid for the triplet state.

$$F_{0\nu_{EX}} \sim \delta_{0\nu_{EX}}. \tag{9.147}$$

Thus, the change of energy of chain 1 is

$$\Delta E_1^{\nu_{EX}} \equiv \Delta E_1^0, \tag{9.148}$$

where ΔE_1^0 is the 0-0 energy difference on chain 1 between the final exciton state and the positive polaron. This is illustrated in Fig. 9.14. By the conservation of energy we therefore have,

$$\Delta E_2^{\nu_{GS}} = -\Delta E_1^0. \tag{9.149}$$

The vibrational level, ν_{GS}, of the ground state of chain 2 to which interconversion from the negative polaron initially occurs is given by

$$
\begin{aligned}
\nu_{GS} &= (\Delta E_2^0 - \Delta E_2^{\nu_{GS}})/\hbar\omega = (\Delta E_2^0 + \Delta E_1^0)/\hbar\omega \\
&= (E_2(P^-;0_{P-}) - E_2(GS;0_{GS}) + E_1(P^+;0_{P+}) - E_1(EX;0_{EX}))/\hbar\omega \\
&= (\Delta E_{CT} - \Delta E_{EX})/\hbar\omega,
\end{aligned}
\tag{9.150}
$$

where

$$\Delta E_{CT} = (E_1(P^+;0_{P+}) - E_1(GS;0_{GS})) + (E_2(P^-;0_{P-}) - E_2(GS;0_{GS})) \tag{9.151}$$

and

$$\Delta E_{EX} = E_1(EX;0_{EX}) - E_1(GS;0_{GS}) \tag{9.152}$$

are the 0-0 transition energies of the charge-transfer exciton and the state $|EX\rangle$, respectively.

9.6.3.3 Multiphonon emission

The condition expressed in eqn (9.148) implies that the energy integral in eqn (9.144) is restricted to the value of $E = \Delta E_1^0$.

Thus, the rate becomes,

$$
\begin{aligned}
k_{I \to F} &= \frac{2\pi}{\hbar} |\langle f|H_\perp|i\rangle|^2 \sum_{\nu_2} F_{0\nu_{GS}}^{(2)} \delta(\Delta E_2^{\nu_{GS}} + \Delta E_1^0) \\
&= \frac{2\pi}{\hbar} |\langle f|H_\perp|i\rangle|^2 F_{0\nu_{GS}}^{(2)} \rho_f(E).
\end{aligned}
\tag{9.153}
$$

$\rho_f(E)$ is the final density of states on chain 2, defined by,

$$\rho_f(E) = \sum_{\nu_{GS}} \delta(\Delta E_2^{\nu_{GS}} + \Delta E_1^0), \tag{9.154}$$

which is usually taken to be the inverse of the vibrational energy spacing.

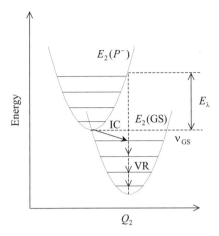

FIG. 9.15. Interconversion (IC) followed by rapid vibrational relaxation (VR) with the emission of ν_{GS} phonons.

Inserting the expression for the Franck Condon factor (see eqn (8.23)),

$$F_{0\nu_{GS}}^{(2)} = |^{(2)}\langle 0_{P^-}|\nu_{1^1 A_g}\rangle^{(2)}|^2 = \frac{\exp(-S_p)S_p^{\nu_{GS}}}{\nu_{GS}!} \tag{9.155}$$

into eqn (9.153) we have the result that,

$$k_{I \to F} = \frac{2\pi}{\hbar}\left|t_\perp \sum_r \psi_{n=1}(r)\tilde{\psi}_{n=1}(r)\right|^2 \rho_f(E)\frac{\exp(-S_p)S_p^{\nu_{GS}}}{\nu_{GS}!}. \tag{9.156}$$

This equation is our final expression for the interconversion rate.

S_p is the Huang-Rhys parameter for the polaron relative to the ground state, defined as

$$E_\lambda/\hbar\omega, \tag{9.157}$$

where E_λ is the reorganization (or relaxation) energy of the polaron relative to the ground state, illustrated in Fig. 9.15. After the isoenergetic transition there is vibrational relaxation to the lowest vibrational level of the ground state of chain 2 via the sequential emission of ν_{GS} phonons. The number of phonons emitted corresponds to the difference in energies between the initial charge-transfer and final exciton states, given by eqn (9.150). This is a multiphonon process, illustrated in Fig. 9.15. In the next section we estimate these rates.

9.6.4 *Estimate of the interconversion rates*

Since interconversion from the intermolecular to the intramolecular charge-transfer excitons is forbidden by symmetry, we now only discuss interconversion to the lowest excitons, S_X or T_X. (As remarked in footnote 44, interconversion to higher-lying exciton states is allowed, but if this happens recombination is an intramolecular process via the intramolecular charge-transfer excitons.) Thus, the state label EX is now taken as S_X or T_X. Similarly, the number of phonons emitted, ν_{GS}, is either ν_S or ν_T, as determined by eqn (9.150) and illustrated in Fig. 9.10.

Within the Mott-Wannier basis the exciton wavefunction overlaps are easy to calculate. Using $t_\perp = 0.1$ eV, the interchain distance as 4 Å and standard semiempirical Coulomb interactions gives

$$\sum_r \psi_{S_X}(r)\tilde{\psi}_{S_{CT}}(r) \approx 1.0 \qquad (9.158)$$

and

$$\sum_r \psi_{T_X}(r)\tilde{\psi}_{T_{CT}}(r) \approx 0.9. \qquad (9.159)$$

The polaron Huang-Rhys parameter, S_p, is not accurately known for light emitting polymers. However, we expect it to be similar to the S_X exciton Huang-Rhys parameter. The relaxation energy of the S_X exciton has been experimentally determined as 0.07 eV in poly(*para*-phenylene vinylene) (Liess *et al.* 1997), with a similar value calculated for 'ladder' poly(*para*-phenylene) (Moore *et al.* 2005).[46] From the figures in Hertel *et al.* (2001), we estimate the relaxation energy to be 0.12 eV in ladder poly(*para*-phenylene) (where the phenyl rings are planar) and 0.25 eV in poly(*para*-phenylene) (where the phenyl rings are free to rotate). Thus, taking the relaxation energy as 0.1 eV and $\hbar\omega = 0.2$ eV implies that $S_p \sim 0.5$.

Now, using $\hbar\omega = 0.2$ eV ($\equiv \rho_f^{-1}$), $S_p \sim 0.5$ and assuming that the energy difference between the strongly bound singlet exciton (S_X) and the *intramolecular* charge-transfer excitons of ~ 0.8 eV is approximately the energy difference between the singlet exciton and the *intermolecular* charge-transfer excitons, we can estimate the interconversion rate for the singlet exciton. This is $k_{S_{CT} \to S_X} \sim 7.5 \times 10^{11}$ s^{-1} (or $\tau_{S_{CT}} \sim 1$ ps). Similarly, using an exchange gap of ~ 0.7 eV gives $k_{T_{CT} \to T_X} \sim 1 \times 10^8$ s^{-1} (or $\tau_{T_{CT}} \sim 10$ ns). Thus, the triplet interconversion rate is much slower than the singlet interconversion rate.

The ISC rate is also not accurately known, with quoted values ranging from 0.3 ns (Burin and Ratner 1998) to 4 ns (Frolov *et al.* 1997). Nevertheless, despite this uncertainty, we see that the estimated triplet interconversion rate is comparable to or slower than the ISC rate, which from eqn (9.112) and Fig. 9.11 implies a large singlet exciton yield

Generally, the ratio of the rates is

$$\frac{k_{S_{CT} \to S_X}}{k_{T_{CT} \to T_X}} = \frac{\left|\sum_r \psi_{S_X}(r)\tilde{\psi}_{S_{CT}}(r)\right|^2 \exp(-S_p)\frac{S_p^{\nu_S}}{\nu_S!}}{\left|\sum_r \psi_{T_X}(r)\tilde{\psi}_{T_{CT}}(r)\right|^2 \exp(-S_p)\frac{S_p^{\nu_T}}{\nu_T!}}. \qquad (9.160)$$

---

[46]Electron-lattice relaxation in light emitting polymers is discussed in Section 11.6.

Thus,

$$\frac{k_{S_{CT} \to S_X}}{k_{T_{CT} \to T_X}} \approx S_p^{-(\nu_T - \nu_S)} \frac{\nu_T!}{\nu_S!}. \tag{9.161}$$

This ratio increases as S_p decreases, because then multiphonon emission becomes more difficult. The ratio also increases as the exchange energy, $(\nu_T - \nu_S)\hbar\omega$, increases for any ν_S or ν_T if $S_p < 1$.

9.6.5 *Discussion and conclusions*

This section has described a theory of intermolecular interconversion from intermolecular weakly bound polaron pairs to strongly bound intramolecular excitons to explain the enhanced singlet exciton yield in light emitting polymers. A crucial aspect of the theory is that both the strongly bound intramolecular excitons and the intermolecular charge-transfer excitons are effective-particles, which are described by both an effective-particle wavefunction and a center-of-mass wavefunction. The orthonormality of the centre-of-mass wavefunctions ensures that interconversion occurs to the lowest member of the strongly bound exciton families, and not to higher lying members of these families. The interconversion is then predominately a multiphonon process, determined by the Franck-Condon factors. These factors are exponentially smaller for the triplet manifold than the singlet manifold because of the large exchange energy. There is also a contribution to the rates from the overlap of the effective-particle wavefunctions, which again are smaller in the triplet manifold, because the triplet exciton has a smaller particle-hole separation and has more covalent character than its singlet counterpart (Tandon *et al.* 2003). As a consequence, the interconversion rate in the triplet manifold is significantly smaller than that of the singlet manifold, and indeed it is comparable to the ISC rates. Thus, the singlet exciton yield is expected to be considerably enhanced over the spin-independent value of 25% in light emitting polymers.

Any successful theory must explain the observation that the singlet exciton yield is close to 25% for molecules and increases with conjugation length (Ho *et al.* 2000; Wohlgenannt *et al.* 2002). This theory qualitatively predicts this trend for two reasons. First, the effective-particle description of the exciton states is only formally exact for infinitely long chains. This description breaks down when the chain length is comparable to the particle-hole separation. In that limit the quantum numbers n and j (which describe the effective-particle wavefunction and center-of-mass wavefunction, respectively) are no longer independent quantum numbers. Then, interconversion is expected to take place between all the states lying between the charge-transfer state and the lowest exciton state. However, as the chain length increases interconversion to higher lying states is suppressed in favour of the lowest lying exciton. This prediction is confirmed by quantum chemical calculations (Beljonne *et al.* 2004). The second reason that the singlet exciton yield is enhanced in polymers over molecules is that as shown in Section 11.6 the Huang-Rhys parameters decrease as the conjugation length increases.

(The experimental evidence for this is discussed by Wohlgenannt *et al.* (2004).)
Thus the relative rate (given by eqn (9.160)) increases.

We note that the effective-particle description is still valid when there is self-trapping. In this case the centre-of-mass wavefuctions are not the particle-in-the-box wavefunctions appropriate for a linear chains (eqn (9.130)), but they are the ortho-normalized functions appropriate for the particular potential well trapping the effective particle. The key point is that because these are ortho-normalized functions and because the potential wells for the excitons and polarons are very similar the selection rules for interconversion are still valid. Thus, interconversion occurs between a pair of states with the same center-of-mass quantum numbers.

Finally, we remark that this theory presents strategies for enhancing the singlet exciton yield. Ideally, the polymer chains should be well conjugated, closely packed and parallel. The last two conditions ensure that the interchain charge-transfer excitons lie energetically below high lying intramolecular excitons, and thus recombination is an interchain interconversion process and not an intramolecular process via the intramolecular charge-transfer excitons. Intramolecular recombination is not desirable because although interconversion from the intrachain charge-transfer excitons is slower in the triplet manifold than the singlet manifold, both rates are expected to be faster than the ISC rate.

10

LINEAR POLYENES AND *TRANS*-POLYACETYLENE

10.1 Introduction

The many unusual properties of *trans*-polyacetylene have already been described in this book in the context of understanding the roles of electron-electron and electron-lattice interactions in π-electron models. These have been described in Chapters 4 and 7, in particular. In this chapter we describe some experimental observations and show how these are explained within a framework of correlated electrons with strong electron-lattice coupling. The chemical structure of *trans*-polyacetylene is illustrated in Fig. 1.1.

The realization that electronic interactions play a significant role in polyenes came via the experimental observations by Hudson and Kohler (1972) that the dipole-forbidden $2^1 A_g$ state lies below the dipole-allowed $1^1 B_u$ state. Further extensive studies of polyene oligomers by Kohler confirm the hypothesis that the *relaxed* energy of the $2^1 A_g$ state lies below the relaxed energy of the $1^1 B_u$ state. An analysis of oligomer spectroscopy from 6 to 16 carbon atoms suggests the empirical relation (Kohler 1988),

$$E^{0-0}(2^1 A_g) = 0.96 + \frac{20.72}{N} \text{ eV,} \tag{10.1}$$

where N is the number of carbon atoms. Similarly, for the $1^1 B_u$ state,

$$E^{0-0}(1^1 B_u) = 2.01 + \frac{15.60}{N} \text{ eV.} \tag{10.2}$$

The existence of the $2^1 A_g$ state below the $1^1 B_u$ state in polyacetylene thin films was suggested by a number of experiments. Third harmonic generation (THG) and two-photon absorption by Halvorson and Heeger (1993) indicates that a $^1 A_g$ state lies below 1.1 eV in energy, while linear absorption, which locates the $1^1 B_u$ state, typically rises at 1.8 eV and peaks at 2.0 eV (Vardeny 1993), as illustrated in Fig. 10.1. Peaks at 0.6 and 0.89 eV were observed in the THG spectrum by Fann *et al.* (1989). From the discussion of THG in Chapter 8, we know that this data implies some combination of $^1 A_g$ states at twice the photon energy (namely 1.2 and 1.8 eV), and $^1 B_u$ states at three times the photon energy (namely 1.8 and 2.7 eV). Fann *et al.* interpreted the experiments as indicating the $^1 A_g$ and $^1 B_u$ states virtually coincident at 1.8 eV.

As we will describe in this Chapter, theoretical modelling indicates that the $2^1 A_g^+$ state, unlike the $1^1 B_u^-$ state, undergoes considerable lattice relaxation. A reasonable interpretation of these experiments is therefore that the relaxed

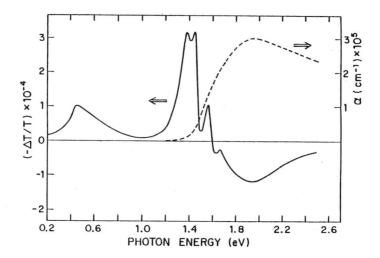

FIG. 10.1. The photoinduced absorption spectrum (solid curve) and linear absorption
(dashed curve) of *trans*-polyactelyene thin film. Reprinted with permission from
Z. V. Vardeny, *Relaxation in Polymers*, edited by T. Kobayashi, World Scientific,
Singapore, 1993. Copyright 1993 by World Scientific Publishing Co. Pte. Ltd.

energy of the $2^1A_g^+$ state lies ca. 1.0 eV below that of the $1^1B_u^-$ state, but that
their vertical transition energies are very similar. However, it is also possible that
the 0.89 eV feature observed in THG spectrum is a higher lying $n^1B_u^-$ state at
2.7 eV. This would be consistent with the interpretation of the $m^1A_g^+$ state at
~ 2.5 eV in the electroabsorption spectrum, as discussed below.

Figure 10.2 illustrates the electroabsorption spectrum of phenyl-substituted
trans-polyacetylene thin film (Liess *et al.* 1997). The feature at 2.0 eV is the red-
shifted 1^1B_u exciton. The feature at 2.5 eV is attributed to a dipole-forbidden
state, namely the m^1A_g state. Unlike polydiacetylene crystals, disordered *trans*-
polyacetylene thin film does not exhibit Franz-Keldysh oscillations (described in
Chapter 8) and therefore a definite assignment of a conduction band edge cannot
made. However, because disordered polydiacetylene also does not exhibit Franz-
Keldysh oscillations, but a smeared-out feature similar to the one exhibited at
2.5 eV in Fig. 10.2 it is sometimes assumed that this feature does mark the band
edge. Another interpretation is that this feature represents the $n = 2$ Mott-
Hubbard exciton,[47] described in Chapter 6, with the particle-hole continuum
lying close in energy (possibly at 2.7 eV, which is three times the THG feature at

---

[47]We note that this is not the $n = 2$ Mott-Wannier exciton, because that is part of the $2^1A_g^+$
state, whose vertical transition energy is expected to lie much closer in energy to the $1^1B_u^-$
state.

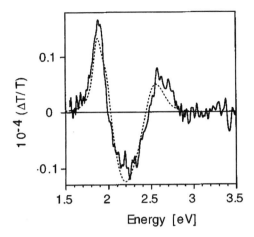

FIG. 10.2. The electroabsorption spectrum of phenyl-substituted *trans*-polyactelyene thin film. Reprinted with permission from M. Liess, S. Jeglinski, Z. V. Vardeny, M. Ozaki, K. Yoshino, Y. Ding, and T. Barton, *Phys. Rev. B* **56**, 15712, 1997. Copyright 1997 by the American Physical Society.

0.89 eV discussed above). Whether the correct interpretation of the 2.5 eV feature is that it is the $n = 2$ Mott-Hubbard exciton or the particle-hole continuum, it seems reasonable to assume that 0.5 eV is a lower bound on the binding energy of the 1^1B_u exciton.

To conclude the experimental review we describe the photoinduced absorption spectrum, which is reproduced in Fig. 10.1. The photoinduced absorption spectrum of a system, obtained while it is being pumped at an energy above the optical gap, gives an insight into the excited states of that system. Typically a polyacetylene system is pumped at 2.4 eV and photoinduced absorption peaks are observed at 0.43 and 1.35 eV. These are referred to as low energy and high energy features, respectively. The low energy feature is attributed to a charged state, as it is associated with infrared modes (Friend *et al.* 1987). In contrast, the high energy feature is attributed to a neutral state, as in this case there are no associated infrared modes. The possible origins of these features will be further discussed in Section 10.2.3.

Theoretical work also suggests the important role of electronic interactions in linear polyenes. By performing a double-configuration-interaction calculation on the Pariser-Parr-Pople model, Schulten and Karplus (1972) demonstrated that the $2^1A_g^+$ state has a strong triplet-triplet contribution, and has a lower energy than the $1^1B_u^-$ state. The triplet-triplet and correlated nature of the $2^1A_g^+$ state was further investigated by Tavan and Schulten (1987). A real-space renormalization group calculation on the Hubbard-Peierls model for chains of up to 16 sites by Hayden and Mele (1986) indicated that the $2^1A_g^+$ state is composed of four solitons. These predictions were confirmed by Su (1995) and Wen and Su (1997). In a different context, Ovchinnikov *et al.* (1973) also highlighted the role of electronic interactions by suggesting that they are largely responsible for the optical gap.

In the next section we describe how the semiempirical Pariser-Parr-Pople-Peierls model - a correlated π-electron model with electron-lattice coupling - quantitatively predicts the excitation spectrum of polyene oligomers, while it qualitatively predicts the spectrum for *trans*-polyacetylene thin films.

10.2 Predictions from the Pariser-Parr-Pople-Peierls model

The predictions of the Pariser-Parr-Pople-Peierls model (defined by eqn (7.1)) as a function of the Coulomb interaction strength for the linear polyene structure were described in Chapter 7. In this section we discuss the predictions taken from Barford *et al.* (2001) of the model for the particular parameter set relevant for *trans*-polyacetylene.

The four parameters in the model are the transfer integral, t, the Coulomb interaction, U, the dimensionless electron-phonon coupling, λ, and α, which relates the bond dimerization Δ_n to the change in bond length. An optimal parametrization of t and U for conjugated polymers was derived by Bursill *et al.* (1998) by fitting the Pariser-Parr-Pople model to the excited states of benzene. This gives $t = 2.539$ eV and $U = 10.06$. Assuming that this parametrization is transferable between all π-conjugated systems it can also be used for *trans*-polyacetylene. λ was found by fitting the vertical energies of the $1^1B_u^-$ and $2^1A_g^+$ states calculated from the Pariser-Parr-Pople-Peierls model to the six-site linear polyene (Bursill and Barford 1999). This gives $\lambda = 0.115$. Finally, using the experimentally determined value of the spring constant, K, as 46 eV $\mathring{A}^{-2}$ (Ehrenfreund *et al.* 1987) gives $\alpha = 4.593$ eV $\mathring{A}^{-1}$.

Solving the Pariser-Parr-Pople-Peierls model with these parameters using the DMRG method and equilibrating using the Hellmann-Feynman method gives a ground state dimerization, $\delta = 0.10$, implying a bond-length alternation of 0.056 $\mathring{A}$. This result is in close agreement with the experimental result of 0.052 $\mathring{A}$ (Kahlert *et al.* 1987).

10.2.1 *Transition energies*

Next, we consider the vertical transition energies, E^v, and the relaxed transition energies, E^{0-0}. These are plotted as a function of inverse chain length in Fig. 10.3(a). We see that the vertical energy of the $2^1A_g^+$ state lies approximately 0.3 eV above that of the $1^1B_u^-$ state in the long chain limit. The relaxation energy of the $1^1B_u^-$ state is modest, being approximately 0.2 eV for 102 sites. By contrast, the relaxation energy of the $1^3B_u^+$ and $2^1A_g^+$ states are substantial, being approximately 0.8 and 1.5 eV, respectively. The energy of the relaxed $2^1A_g^+$ state lies 1 eV below that of the $1^1B_u^-$ state. As already described in Chapter

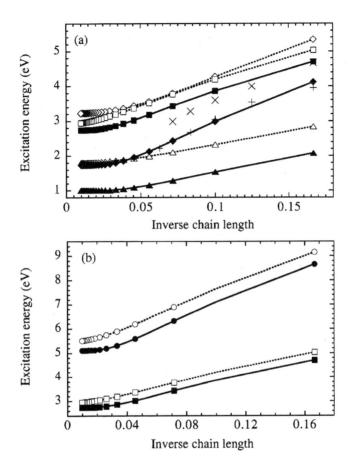

FIG. 10.3. (a) Calculated transition energies for the $1^1B_u^-$ state (squares), $2^1A_g^+$ state (diamonds) and $1^3B_u^+$ state (triangles) as functions of the inverse number of sites. Vertical and relaxed transitions are indicated by dashed and solid lines and open and solid symbols, respectively. Experimental values of the relaxed $1^1B_u^-$ ($\times$) and $2^1A_g^+$ (+) state energies for polyenes in hydrocarbon solution (Kohler 1988). (b) Calculated transition energies for the $1^1B_u^-$ state (squares) and charge gap (circles). Reprinted with permission from W. Barford, R. J. Bursill, and M. Yu Lavrentiev, *Phys. Rev. B* **63**, 195108, 2001. Copyright 2001 by the American Physical Society.

7, the strong relaxation of the $2^1A_g^+$ state is associated with a large distortion from the ground state structure.

The experimental values of the $E^{0-0}(1^1B_u^-)$ and $E^{0-0}(2^1A_g^+)$ for short polyenes are also shown (Kohler 1988). The calculated $2^1A_g^+$ results are in excellent agreement with the experimental values. The $1^1B_u^-$ values are approximately 0.3 eV lower than the theoretical predictions, which is approximately the reduction expected by the solvation of the polyene chain in solution, as discussed in Section 9.4. In contrast, since the $2^1A_g^+$ state is a highly correlated state with more spin-density-wave (or covalent) character than particle-hole (or ionic) character, this state is not expected to exhibit much solvation (Barford *et al.* 2004).

However, while the calculated values for short chains fit the Kohler empirical expressions (eqns (10.1) and (10.2)) rather well, there are significant deviations from them for long chains, and as a consequence the calculated values also deviate from the thin film experimental values for the $1^1B_u^-$ and $2^1A_g^+$ states described in Section 10.1.

The calculated energies converge rapidly with chain size because of self-trapping, which occurs once the chain size exceeds the spatial extent of the solitonic structures. These will be described in the next section. In Section 10.3 we discuss the extent to which a fully quantum treatment of the lattice degrees of freedom changes this behaviour to better fit the experimental values.

Figure 10.3(b) shows the the charge gap, 2Δ, defined as

$$2\Delta = E_0(N + 1) + E_0(N - 1) - 2E_0(N), \qquad (10.3)$$

where $E_0(N)$ is the ground state energy for N electrons. It also shows the transition energy of the $1^1B_u^-$ state. In the long chain limit the charge gap represents the energy of an uncorrelated electron-hole pair, and therefore represents the band edge. The relaxation energy of the charge gap is roughly double that of the $1^1B_u^-$ state. This is expected as the two charges form independent polarons, whereas the excitonic $1^1B_u^-$ state forms a single polaron, as described in Chapter 7.

The isolated single chain binding energy is 2.4 eV. However, as explained in Section 9.4, the unbound pair is strongly screened in a solid state environment by ~ 2.0 eV, whereas the exciton is more weakly screened (~ 0.3 eV). This implies that the binding energy of the exciton in the solid state is ~ 0.7 eV, in reasonable agreement with experimental interpretations of Section 10.1.

10.2.2 *Soliton structures*

The structures of the ground state and the $1^1B_u^-$, $1^3B_u^+$ and $2^1A_g^+$ states are shown in Fig. 10.4. The $1^3B_u^+$ and $2^1A_g^+$ states undergo considerable lattice distortion, whereas the 1^1B_u state shows a weak polaronic distortion of the lattice, very similar to the charged state. In Chapter 7 it was shown that the $1^1B_u^-$ and $2^1A_g^+$ states fit a two-soliton form (defined in eqn (7.19)). In contrast, the $2^1A_g^+$ state fits a four-soliton form (defined in eqn (7.20)), indicating the strong triplet-triplet character of that state.

Further insight into the electronic structure of polyenes and its relation to their geometry can be obtained from the spin-spin correlation function, defined as

$$S_n = \langle \hat{S}_n^z \hat{S}_{N+1-n}^z \rangle, \qquad (10.4)$$

where $\hat{S}_n^z$ is the z-component of the spin operator on site n. S_n measures antiferromagnetic correlations between sites symmetrically situated with respect to the

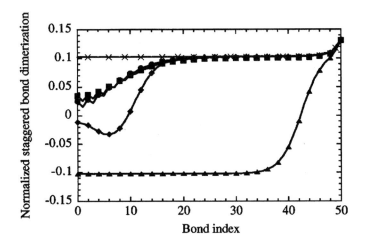

FIG. 10.4. The staggered, normalized bond dimerization, δ_n, (defined in eqn (4.27)) as a function of bond index from the centre of the chain of various states of *trans*-polyacetylene calculated from the Pariser-Parr-Pople-Peierls model. $1^1A_g^+$ (crosses), $1^1B_u^-$ (squares), $1^3B_u^+$ (triangles), $2^1A_g^+$ (diamonds) and polaron (circles). Reprinted with permission from W. Barford, R. J. Bursill, and M. Yu Lavrentiev, *Phys. Rev. B* **63**, 195108, 2001. Copyright 2001 by the American Physical Society.

centre of the chain. As the correlation function shows unimportant oscillations between even and odd site indices it is more convenient to use the symmetrized function,

$$\tilde{S}_m = \frac{1}{2}\left(S_{(N-m)/2} + S_{(N-m)/2+1}\right), \qquad (10.5)$$

where $m = 0, 4, 8, \ldots, N-2$. This measures the correlations between pairs of doubly bonded sites, with m being the distance between them.

The spin-spin correlation functions calculated in the ground state geometry are shown in Fig. 10.5(a). They show a monotonic decay for the correlations in the $1^1A_g^+$ and $1^1B_u^-$ states, but in the $2^1A_g^+$ state there is a small minimum at $m = 8$ and a maximum at $m = 16$. This behaviour of the spin-spin correlations in the $2^1A_g^+$ state becomes clearer when it is calculated in the relaxed geometry of this state. Here, the correlation function of the $2^1A_g^+$ state, shown in Fig. 10.5(b), has a strong minimum at $m = 8$ - where it changes sign - and a maximum at $m = 20$. These features strongly confirm the triplet-triplet character of this state. By comparing Fig. 10.5(b) to the soliton structure shown in Fig. 10.4 we see that the unpaired spins (or spinons) correspond to the positions of the four geometrical solitons at $n = \pm m/2$, namely $n = \pm 4$ and $n = \pm 10$.

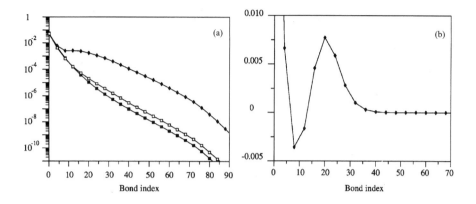

FIG. 10.5. Spin-spin correlation function (defined in eqn (10.5)) as a function of the distance *between* the pairs of spins for the $1^1A_g^+$ (solid squares), $2^1A_g^+$ (solid diamonds) and $1^1B_u^-$ (empty squares) states. (a) In the relaxed ground state geometry. (b) In the relaxed $2^1A_g^+$ geometry. Reprinted with permission from W. Barford, R. J. Bursill, and M. Yu Lavrentiev, *Phys. Rev. B* **63**, 195108, 2001. Copyright 2001 by the American Physical Society.

10.2.3   *Adiabatic potential energy curves*

The soliton-antisoliton interactions in the excited states are illustrated by the adiabatic energy curves shown in Fig. 10.6. These are obtained by calculating transition energies as a function of the soliton-antisoliton separation, $2n_0$, using the equilibrium values of ξ (and n_d/n_0 for the $2^1A_g^+$ state). The two-soliton fit (eqn (7.19)) is used for the $1^1B_u^-$ state for a fixed value of $\xi = 11.8$. The four-soliton fit (eqn ((7.20))) is used for the $2^1A_g^+$ state for fixed values of $\xi = 5.7$ and $n_d/n_0 = 1.5$.

Figure 10.6 shows a rather weak repulsion for the soliton pairs in the $1^1B_u^-$ state at short distances, leading to the shallow polaronic distortion, as already discussed. At longer distances there is strong attraction, arising from the excitonic character of this state. In contrast, for the $2^1A_g^+$ state there is a strong short range repulsion and a weak long range attraction from its residual particle-hole character.

Figure 10.6 illustrates the crossover in energies of the $1^1B_u^-$ and $2^1A_g^+$ states as a function of soliton separation. Thus, a vertical photo-excitation to the 1^1B_u state would be rapidly followed by the creation of a soliton-antisoliton pair. If symmetry breaking interconversion interactions are present there will be a crossover to the $2^1A_g^+$ state. Since this state is dipole inactive emission to the ground state is nonradiative.

Figure 10.6 also provides a possible explanation of the high energy feature observed at 1.35 eV in the photo-induced absorption spectrum of *trans*-polyacetylene, which is attributed to neutral states. A vertical excitation from

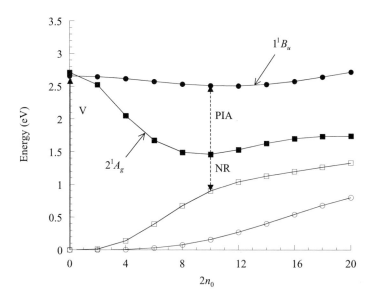

FIG. 10.6. The adiabatic potential energy curves for the $1^1B_u^-$ state (solid circles) and the $2^1A_g^+$ state (solid squares) as a function of the soliton-antisoliton separation, $2n_0$. The corresponding energy of the ground state in the geometry of the relaxed excited states are shown in open symbols. The vertical optical transition from the ground state to the $1^1B_u^-$ state is denoted by a solid up-arrow labelled 'V'; the vertical nonradiative emission from the $2^1A_g^+$ state to the ground state is shown by a dashed down-arrow labelled 'NR'; and the possible origin of the high energy feature in the photoinduced absorption is shown by a dashed up-arrow labelled 'PIA'. The parameters used in this calculation are $t = 2.5$ eV, $U = 10$ eV and $\lambda = 0.1$.

the relaxed energy of the $2^1A_g^+$ state to the $1^1B_u^-$ state is at 1.2 eV, very close in energy to the experimental feature. However, a difficulty with this explanation is that the oscillator strength for this transition is very small (Ramasesha and Soos 1984; Barford *et al.* 2001). An alternative explanation for the high energy feature is that it corresponds to a transition from the lowest triplet state ($1^3B_u^+$) to the $1^3A_g^-$ state. (Indeed, a similar feature in light emitting polymers is attributed to this transition.) This transition does have the required oscillator strength.

The attribution of the low energy feature at 0.43 eV to charged states is less problematic, as it is consistent with the theoretically calculated energy difference of 0.45 eV between the relaxed polaron and its first dipole connected excited state. This transition also has a significant oscillator strength.

10.3   Quantum phonons

The adiabatic approximation is widely accepted as being applicable to the electronic states of conjugated polymers. As described above, solutions of an adiabatic Hamiltonian (namely, the Pariser-Parr-Pople-Peierls model) agree remarkably well with experimental observations for short polyenes. A linear extrapolation in inverse chain length of the experimental observations coincide with the experimental observations of the energies of the $1^1B_u^-$ and $2^1A_g^+$ states in thin films. Thus, it is reasonable to assume that the thin film observations correspond to excitations in well-conjugated polymers. In contrast, the calculated excitation energies deviate from a linear extrapolation once self-trapping of the excited states occurs. This occurs when the chain length exceeds the spatial extent of the soliton structures. Thus, it is reasonable to ask, is self-trapping an unphysical artefact of the adiabatic approximation in linear polyenes? Correspondingly, will a fully quantized model achieve better predictions for the excitation energies? These questions were addressed by Barford et $al.$ (2002a). We reproduce that analysis here.

The fully quantized model is described by the following Hamiltonian,

$$H = H_{\mathrm{ph}} + H_e + H_{e-\mathrm{ph}}. \tag{10.6}$$

The phonon Hamiltonian is,

$$H_{\mathrm{ph}} = \sum_{n=1}^{N} \frac{P_n^2}{2M} + \frac{K}{2} \sum_{n=1}^{N-1} (u_{n+1} - u_n)^2 , \tag{10.7}$$

where M is the nuclear mass and K is the elastic spring constant. Notice that since we are considering linear chains the sum in the elastic term is from $n = 1, \ldots, N-1$.

The electron Hamiltonian is the usual Pariser-Parr-Pople model,

$$
H_e = -t \sum_{n=1,\sigma}^{N-1} (c_{n\sigma}^\dagger c_{n+1\sigma} + c_{n+1\sigma}^\dagger c_{n\sigma}) + U \sum_{n=1}^{N} \left(N_{n\uparrow} - \frac{1}{2}\right)\left(N_{n\downarrow} - \frac{1}{2}\right)
$$
$$
+ \frac{1}{2} \sum_{m \neq n} V_{mn}(N_m - 1)(N_n - 1). \tag{10.8}
$$

Finally, assuming that electron-phonon coupling arises from linear deviations in bond lengths, the electron-phonon coupling Hamiltonian is the same as that introduced in Chapter 7, namely,

$$
H_{e-\mathrm{ph}} = \alpha t \sum_{n\sigma} (u_{n+1} - u_n)(c_{n\sigma}^\dagger c_{n+1\sigma} + c_{n+1\sigma}^\dagger c_{n\sigma})
$$
$$
- 2\alpha W \sum_{n} (u_{n+1} - u_n)(N_{n+1} - 1)(N_n - 1), \tag{10.9}
$$

where W is defined in eqn (7.6).

The Pariser-Parr-Pople-Peierls model is the adiabatic limit of this model, taken by setting $M \to \infty$ and treating the nuclear displacements classically. However, now we intend to quantize the nuclear degrees of freedom. To do this we rewrite H_{ph} as,

$$H_{ph} = \sum_{n=2}^{N-1} \frac{P_n^2}{2M} + \frac{\tilde{K}}{2} \sum_{n=2}^{N-1} u_n^2 - \frac{\tilde{K}}{2} \sum_{n=2}^{N-2} u_{n+1} u_n, \tag{10.10}$$

where $\tilde{K} = 2K$. To ensure constant chain lengths we have also taken the first and last sites to be stationary (thus, $u_1 = u_N = 0$). The first two terms on the right-hand side describe $N-2$ independent harmonic oscillators, while the final term represents the coupling between these oscillators.

The first two terms are diagonalized by introducing the phonon creation and annihilation operators (see (Cohen-Tannoudji *et al.* 1977)):

$$b_n^{\dagger} = \sqrt{\frac{M\omega}{2\hbar}} u_n - i\sqrt{\frac{1}{2M\hbar\omega}} P_n \tag{10.11}$$

and

$$b_n = \sqrt{\frac{M\omega}{2\hbar}} u_n + i\sqrt{\frac{1}{2M\hbar\omega}} P_n, \tag{10.12}$$

respectively. $b_n^{\dagger}$ (b_n) creates (destroys) a quantum of energy, $\hbar\omega$, in the linear harmonic oscillator located on the site n, where $\omega = \sqrt{\tilde{K}/M}$.

The inverse expressions are,

$$u_n = \sqrt{\frac{\hbar}{2M\omega}} \left(b_n^{\dagger} + b_n\right) \tag{10.13}$$

and

$$P_n = i\sqrt{\frac{M\hbar\omega}{2}} \left(b_n^{\dagger} - b_n\right). \tag{10.14}$$

Substituting for u_n and P_n in eqn (10.10) we have,

$$H_{ph} = \hbar\omega \sum_n \left(b_n^{\dagger} b_n + \frac{1}{2}\right) - \hbar\omega \sum_n B_{n+1} B_n, \tag{10.15}$$

where we have introduced,

$$B_n = \frac{\left(b_n^{\dagger} + b_n\right)}{2}, \tag{10.16}$$

which represents the dimensionless displacement of the nth. oscillator.

Similarly,

$$H_{e-\mathrm{ph}} = g \sum_{n\sigma} (B_{n+1} - B_n) \left(c^{\dagger}_{n+1\sigma} c_{n\sigma} + c^{\dagger}_{n\sigma} c_{n+1\sigma} \right)$$
$$- \tilde{W} \sum_n (B_{n+1} - B_n)(N_n - 1)(N_{n+1} - 1), \qquad (10.17)$$

where

$$g = \left(\frac{t\lambda\pi\hbar\omega}{2} \right)^{1/2}, \qquad (10.18)$$

$$\tilde{W} = \left(\frac{\hbar\omega}{K} \right)^{1/2} \frac{U\gamma r_0}{(1 + \gamma r_0^2)^{3/2}}, \qquad (10.19)$$

$\gamma = (U/14.397)^2$ and r_0 is the undistorted average bond length.

Equations (10.8), (10.15), and (10.17), along with the boundary conditions that

$$B_1 = B_N = 0 \qquad (10.20)$$

complete the description of the model. For parameter ranges applicable for linear polyenes, namely $\omega_0 = \sqrt{K/M} = 0.2$ eV, and t, U, λ and α as defined in Section 10.2, this model can be conveniently solved using the density matrix renormalization group method. Full details of the implementation are given in Barford et al. (2002a). There is also a brief description of the methodology in Appendix H. The next section describes the results.

10.3.1   Results and discussion

Figure 10.7 shows the transition energies for the $1^1B_u^-$ and $1^3B_u^+$ states as a function of inverse chain length for up to 102 sites. For short chains the differences between the transition energies in the quantum and adiabatic limits are very small. However, the quantum calculation of the triplet state energy deviates from the adiabatic result in two ways. First, the gradient as a function of inverse chain length is greater, and second, the flattening-off of the energy occurs at a larger chain length. As a consequence, there is a clear deviation between these limits for the triplet state as the conjugation length increases. This deviation is a result of the de-pinning of the excited state by the lattice fluctuations. In contrast, the deviation between the adiabatic and quantum limits for the singlet excited state is relatively modest. At 102 sites the deviations are 0.38 eV and 0.14 eV for the triplet and singlet states, respectively.

Further insight into the de-pinning of the excited states can be obtained from a study of their geometrical structures. Figure 10.8 shows the staggered bond length changes in the ground state and excited states. We see that the ground state dimerization of ca. 0.04 Å in the quantum limit is slightly smaller than the adiabatic result of ca. 0.05 Å. In the limit of long chains the relative root-mean-square fluctuation in the bond length is ca. 0.9, close to previous theoretical (Su 1982) and experimental (McKenzie and Wilkin 1992) estimates. However,

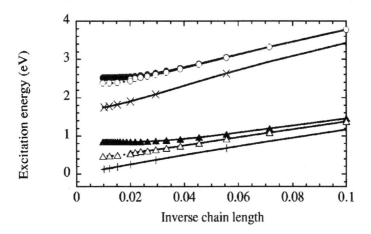

FIG. 10.7. Transition energies for the $1^1B_u^-$ state (circles) and $1^3B_u^+$ state (triangles) as a function of inverse chain length. Adiabatic/quantum calculations are indicated by solid/open symbols. Also shown are the $1^1B_u^-$ ($\times$) and $1^3B_u^+$ (+) transition energies for the undimerized Pariser-Parr-Pople model. Reprinted with permission from W. Barford, R. J. Bursill, and M. Yu Lavrentiev, *Phys. Rev. B* **65**, 75107, 2002. Copyright 2002 by the American Physical Society.

although there are considerable quantum fluctuations in the bond alternation, there is no evidence that these fluctuations destroy the broken symmetry of the ground state.

There are significant deviations between the quantum and adiabatic predictions for the triplet soliton structures. The soliton width in the adiabatic calculation (ca. 10 bond lengths) is much shorter than the corresponding quantum calculation.

Finally, we consider the optically allowed excitonic ($1^1B_u^-$) state. In the adiabatic approximation this state creates a shallow polaronic distortion of the lattice, with self-trapping only becoming important for chain lengths longer than ca. 40 sites. This is confirmed by the excitation energies shown in Fig. 10.7, indicating that the transition energies calculated in the quantum limit are within ca. 0.1 eV of the adiabatic result, and Fig. 10.8, showing that the quantum and adiabatic polaronic structures are similar.

Evidently, quantum lattice fluctuations play an important role in the depinning of the self-trapped excited states, leading to corrections to the adiabatic approximation. These corrections are particularly important for the lowest-lying triplet, as this state is gapless in the long chain limit in the absence of electron-phonon coupling. Figure 10.7 show the triplet transition energy for the undimerized Pariser-Parr-Pople model. Thus, the phonon frequency (0.2 eV) is not small

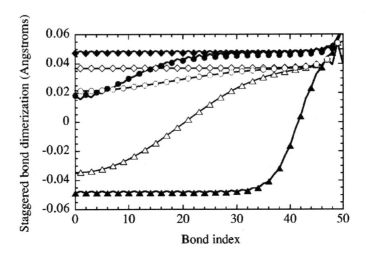

FIG. 10.8. The staggered bond dimerization in $\overset{\circ}{A}$ as a function of bond index from the centre of the chain) of various states: $1^1A_g^+$ (diamonds), $1^1B_u^-$ (circles) and $1^3B_u^+$ (triangles). Adiabatic/quantum calculations are indicated by solid/open symbols. Reprinted with permission from W. Barford, R. J. Bursill, and M. Yu Lavrentiev, *Phys. Rev. B* **65**, 75107, 2002. Copyright 2002 by the American Physical Society.

in comparison to the electronic energy scale, and the approximation of slow nuclear motion relative to the electronic timescales is no longer valid. This breakdown of the adiabatic approximation is an emergent property of long chains. At 102 sites the phonon-calculated triplet energy is only 56% of the adiabatic approximation. Since the dipole forbidden $2^1A_g^+$ state in *trans*-polyacetylene is predominately formed from a pair of bound triplets, this reduction in the triplet energy from quantum fluctuations is also expected to apply to the $2^1A_g^+$ state. It would be reasonable to expect that the semiclassical prediction of 1.74 eV for its transition energy might be reduced to ca. 1.0 eV with the inclusion of quantum phonons. This prediction is very close to Kohler's linear extrapolation (1988), and to the experimental determination of the $2^1A_g^+$ energy by Halverson and Heeger (1993).

In contrast, the exciton-polaron ($1^1B_u^-$) state is expected to be in the adiabatic limit, as its energy in the undimerized Pariser-Parr-Pople model is 1.6 eV in the long chain limit. This is confirmed by Fig. 10.7, which show the deviations between the quantum and adiabatic limits is only ca. 0.1 eV at 102 sites.

We have seen in this section how quantum fluctuations can reconcile the predictions of π-electron models to the experimental observations on thin films. However, remembering that disorder is also an effective mechanism to pin excited states, it is possible that the parametrization of the π-electron models (derived with short oligomers) is simply not valid for long polymers.

10.4   Character of the excited states of *trans*-polyacetylene

The physical parameter range relevant for *trans*-polyacetylene, namely Coulomb interactions comparable to the bandwidth, is intermediate between the weak and strong coupling limits defined in Chapter 5. As shown in Fig. 5.2, this means (as already discussed) that the $1^1B_u^-$ and $2^1A_g^+$ vertical transition energies are virtually degenerate, signalling a bimagnon component to the $2^1A_g^+$ state. Further evidence for this bimagnon character is the four-soliton fit to the geometrical structure. However, there is still some residual $n = 2$ Mott-Wannier exciton character to the $2^1A_g^+$ state, because as shown in Fig. 10.6 the solitons are weakly attracted at large separation. This is in contrast to the spin-1/2 spinons of the $1^3B_u^+$ state, which are weakly repelling at all separations. (In the strong coupling limit the $2^1A_g^+$ state evolves to a pair of unbound magnons, as described in Chapter 7.)

The character of the $1^1B_u^-$ and $1^3B_u^+$ states are easier to understand, and have already been described in Chapters 6 and 7. The $1^1B_u^-$ state in the intermediate regime has both $n = 1$ Mott-Wannier exciton and $n = 1$ Mott-Hubbard exciton character. Electron-lattice coupling enhances this particle-hole character, with the exciton being composed of a pair of charged-spinless solitons ($S^\pm$). The $1^3B_u^+$ state in the intermediate regime has both $n = 1$ Mott-Wannier exciton and spin-density-wave character. Electron-lattice coupling enhances the 'covalent' character causing a pair of spin-1/2 spinons (S^σ). The $1^1B_u^-$ and $1^3B_u^+$ states are schematically illustrated in Fig. 7.7.

Finally, we remark on the m^1A_g state shown in the electroabsorption spectrum of Fig. 10.2. While it is possible that 2.5 eV is the vertical transition energy of the $2^1A_g^+$ state, the THG experiments of Fann *et al.* (1989) indicate that the vertical transition energies of the $2^1A_g^+$ and $1^1B_u^-$ states are virtually degenerate. Thus, more reasonable interpretations are that the 2.5 eV feature represents either the $n = 2$ Mott-Hubbard exciton or the particle-hole continuum

10.5   Other theoretical approaches

A recent *ab initio* calculation of the optical spectrum of *trans*-polyacetylene has been performed by Rohlfing and Louie (1999). Their approach is to correct the quasi-particle gap obtained within density functional theory by the GW-approximation and then to construct a Bethe-Salpeter equation for the particle-hole excitations. As discussed in Section 6.2.4, this is a weak-coupling approximation, as it assumes the existence of valence and conduction band quasi-particles. Although the calculation was performed for a single chain, the effects of bulk dielectric screening were modelled by a screened particle-hole interaction. Figure 10.9(a) shows the calculated optical spectrum, with a predicted binding energy of 0.4 eV for the lowest exciton. Figure 10.9(b) shows the exciton probability distribution functions for the zero-momentum first and second excited states, namely the $n = 1$ (even parity) and $n = 2$ (odd parity) Mott-Wannier excitons. These are equivalent to the $1^1B_u^-$ and $2^1A_g^+$ states for linear systems. The near degeneracy of these energies is in agreement with the Pariser-Parr-Pople-Peierls

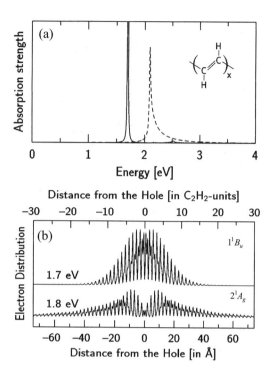

FIG. 10.9. (a) Calculated optical absorption spectrum of *trans*-polyacetylene from a DFT-GWA-BSE calculation. The solid and dashed curves represent the exciton and quasi-particle spectra, respectively. (b) The electron-hole distribution function. Reprinted with permission from M. Rohlfing and S. G. Louie, *Phys. Rev. Lett.*, **82**, 1959, 1999. Copyright 1999 by the American Physical Society.

model calculations and some experiments. Notice, however, that this procedure cannot describe Mott-Hubbard excitons, so the possibility of a Mott-Hubbard exciton at 2.5 eV has not been demonstrated.

11

LIGHT EMITTING POLYMERS

11.1 Introduction

The discovery of electroluminescence in poly(*para*-phenylene vinylene) (PPV) (Burroughes *et al.* 1990) has led to a re-awakened interest in conjugated polymers. This interest is partly driven by a desire to understand the electronic properties of the phenyl-based light emitting polymers in order to exploit them for a wide range of technologies. These technologies include cheap and flexible light emitting displays, photovoltaic devices, optical switching, and field-effect transistors.

In this chapter we present a description of the excited states of the phenyl-based light emitting polymers.[48] This description is achieved by using the theoretical and computational modelling of these systems to interpret the experimental evidence acquired by a wide variety of spectroscopic probes.

The first observation we make is that, in contrast to linear polyenes, the phenyl-based polymers electroluminesce. This indicates that the electronic states are different in the two systems. In particular, although electronic interactions are strong in phenyl-based systems, they are not as strong as in linear polyenes to cause the reversal in energetic ordering of the dipole-allowed and dipole-forbidden singlets that is observed in the latter systems. There are two qualitative explanations why electronic correlations are less strong in phenyl-based systems than in linear polyenes. First, the presence of phenyl rings in the chemical structure means that electrons are more able to avoid each other than in a linear chain. This means that a mapping of the low-energy physics onto an effective one-dimensional model would imply a reduced Coulomb interaction, U. Second, the mapping of the valence and conduction band structures onto an effective one-dimensional model implies a relatively large effective bond alternation, namely $\delta \sim 0.2$, in contrast to $\delta \sim 0.1$ in *trans*-polyacetylene (Soos *et al.* 1993). As described in Chapter 5, a larger effective bond alternation implies reduced electronic correlations. As a consequence, the low energy excitations of light emitting polymers (namely those excitations associated with peak I of the absorption spectrum described below) may be described by the weak-coupling exciton theory introduced in Chapters 5 and 6. However, these arguments concerning electronic correlations do not apply to the higher lying excitations (namely those excitations associated with peaks II, III, and IV of the absorption spectrum described below), as these excitations are highly localized.

---

[48]See Section 9.6 for a discussion of the singlet exciton yield in light emitting polymers.

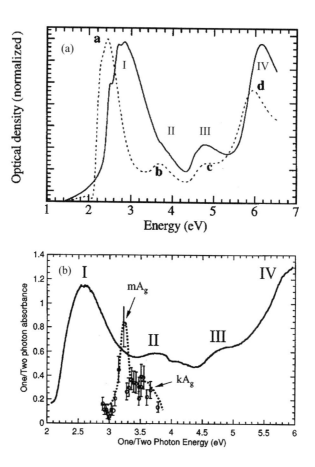

Fig. 11.1. (a) The optical-absorption spectra of PPV (solid curve) and MEH-PPV (dashed curve). Reprinted with permission from S. J. Martin, D. D. C. Bradley, P. A. Lane, H. Mellor, and P. L. Burn, *Phys. Rev. B* **59**, 15133, 1999. Copyright 1999 by the American Physical Society. (b) The optical-absorption (one-photon) spectra of a DOO-PPV film (solid curve) and two-photon absorption of DOO-PPV in solution (circles). Reprinted with permission from S. Frolov, Z. Bao, M. Wohlgenannt, and Z. V. Vardeny, *Phys. Rev. B* **65**, 205209, 2002. Copyright 2002 by the American Physical Society. The chemical structures of PPV, MEH-PPV and DOO-PPV are shown in Fig. 11.3.

We now turn to a description of the optical properties. Figures 11.1 and 11.2 show the characteristic linear absorption spectrum of the phenyl-based light emitting polymers. These are

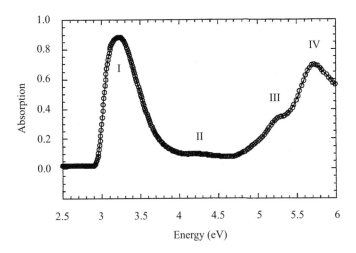

FIG. 11.2. The linear-absorption of PFO (Cadby and Martin 2004). The chemical structure of PFO is shown in Fig. 11.3.

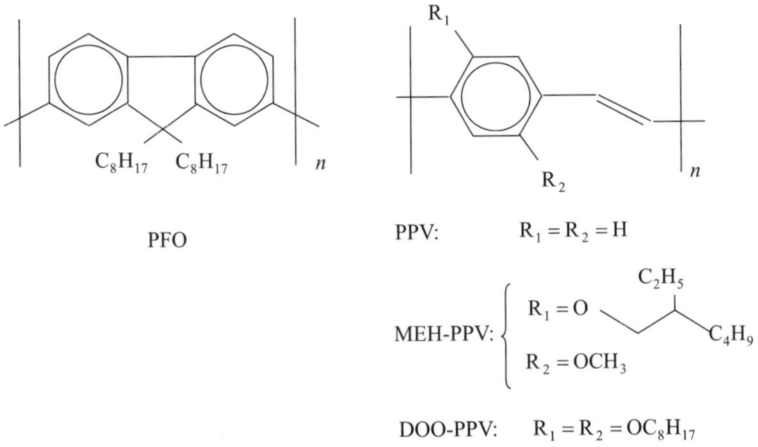

PFO

PPV: $R_1 = R_2 = H$

MEH-PPV: $\left\{ \begin{array}{l} R_1 = O \\ R_2 = OCH_3 \end{array} \right.$

DOO-PPV: $R_1 = R_2 = OC_8H_{17}$

FIG. 11.3. The chemical structures of PFO, PPV, MEH-PPV and DOO-PPV.

- A dominant low energy peak (labelled I or a), predominantly polarized along the long-axis. This is the 1^1B_u state.

- A weak peak (labelled II or b), predominantly polarized along the short-axis in poly(*para*-phenylene) (PPP), but along the long-axis in PPV. This feature becomes more pronounced when there is chemical substitution, as in the case of MEH-PPV illustrated in Fig. 11.1(a).

- Another relatively weak peak (labelled III or c), predominantly polarized along the short-axis.

- A dominant high energy peak (labelled IV or d), polarized along the long-axis.

As well as linear spectroscopy, the nonlinear spectroscopic techniques of electroabsorption, third harmonic generation and two-photon absorption have all been deployed to investigate other excited states. In addition, phosphorescence probes and photoinduced absorption have been used to investigate the triplet states. In view of the spectral shifts arising from disorder and variations in the chemical structures, these investigations reveal a remarkably consistent picture for PPP, PFO, and PPV.

In particular,

- Electroabsorption (Martin *et al.* 1999), and two-photon absorption and photoinduced absorption (Frolov *et al.* 2002) indicate a dipole forbidden state at ca. 0.7 eV higher in energy than the 1^1B_u state in PPV derivatives. This state is labelled the m^1A_g state, and is indicated in Fig. 11.1(b). Electroabsorption in PFO indicates that the m^1A_g state is 0.8 eV higher in energy than the 1^1B_u state (Cadby *et al.* 2000).

- Third harmonic generation indicates a ^1B_u state at 3.2 eV in PPV (Mathy *et al.* 1996). Modelling of the electroabsorption data by Martin *et al.* (1999) indicates that this state is ca. 0.1 eV higher in energy than the m^1A_g state. This ^1B_u state is labelled the n^1B_u state.

- Phosphorescence indicates a triplet state at ca. 0.7 eV lower in energy than the 1^1B_u state for a wide variety of systems. (See Köhler and Beljonne (2004) for a review of the data.) This triplet state is the 1^3B_u state.

- Photoinduced absorption from the 1^3B_u state indicates another triplet state 1.4 eV higher in energy (Monkman *et al.* 2001). This state is the 1^3A_g state, which is almost degenerate with the m^1A_g state.

- Photoinduced absorption and two-photon absorption (Frolov *et al.* 2002) indicates another dipole forbidden state at 3.6 eV in PPV. This long-lived state is labelled k^1A_g in Fig. 11.1(b). A strong photoinduced absorption signal has also been observed at 1.5 eV above the relaxed 1^1B_u state in PFO (Xu *et al.* 2001).

- Photoconduction in MEH-PPV occurs at 3.1 eV (Chandross *et al.* 1994), while in ladder-type PPP it occurs at 1.1 eV above the 0-0 transition to the 1^1B_u state (Barth *et al.* 1998).

Although the transition energies of peaks I-III vary between different kinds of phenyl-based light emitting polymers, the observation that the general spectroscopic features, and even the actual energy gaps between excited states are so similar between different systems, suggests that a common description exists for the excited states. We will argue in this chapter that the 'low-energy' states

Table 11.1 *The spectroscopically determined state energies in eV*

Polymer	1^1B_u	m^1A_g	n^1B_u	1^1B_u Exciton binding energy
PPV[a]	2.84	—	—	—
PPV[b]	2.46	3.15	3.3	0.84
PPV[c]	—	—	3.2	—
MEH-PPV[d]	2.44	—	—	—
MEH-PPV[e]	2.25	2.9	3.0	0.75
DOO-PPV[f]	2.5	3.2	—	> 0.7
PFO[g]	3.2	4.0	—	> 0.8
PPP[h]	3.7	4.6	—	> 0.9

[a]Vertical excitation from linear absorption (Martin *et al.* 1999). [b]Sum-over-states fitting of electroabsorption (Martin *et al.* 1999). [c]Third harmonic generation (Mathy *et al.* 1996). [d]Vertical excitation from linear absorption (Martin *et al.* 1999). [e]Sum-over-states fitting of electroabsorption (Martin *et al.* 1999). [f]Vertical excitations from linear and two-photon absorption (Frolov *et al.* 2002). [g]1^1B_u vertical excitation from linear absorption, m^1A_g from electroabsorption (Cadby *et al.* 2000). [h]1^1B_u vertical excitation from linear absorption, m^1A_g from electroabsorption (Lane *et al.* 1997). The 1^1B_u exciton binding energy is determined by assuming that the n^1B_u state lies at or close to the particle-hole continuum. The m^1A_g energy therefore provides a lower bound to the 1^1B_u exciton binding energy.

(1^1B_u, m^1A_g, n^1B_u, 1^3B_u and 1^3A_g) are all associated with particle-hole excitations from the valence to the conduction band.

Table 11.1 lists the energies of the low energy states as determined by various spectroscopic probes. Later in this chapter we will argue that the 1^1B_u and m^1A_g states are the $n = 1$ and $n = 2$ Mott-Wannier excitons, respectively.[49] The n^1B_u state is expected to be the $n = 3$ Mott-Wannier exciton, lying close to the particle-hole continuum, or the onset of the particle-hole continuum itself. The DMRG calculations presented in Section 11.2.3 suggest the latter possibility. This interpretation of the excited states then places a lower bound on the 1^1B_u exciton binding energy, as listed in the table.

In attempting to understand the excited states of light emitting polymers the theoretical community has often taken two opposing points of view. On the one hand there is the view that the excited states of polymers are derived from those of benzene (Rice and Gartstein 1994; Gartstein *et al.* 1995). Since electronic interactions are important in benzene, this view proposes that the excited states of polymers also exhibit strong electronic correlations. An alternative assumption is that because excited states in polymers are more delocalized than in molecules electronic interactions are less important, and therefore a conventional semiconductor viewpoint of bound electron-hole excitations describes the physics. This viewpoint is strongly advocated by Kirova and Brazovskii (Kirova *et al.* 1999;

---

[49]Mott-Wannier excitons were described in Chapter 6. Recall that our definition of Mott-Wannier excitons includes bound particle-hole excitations with small particle-hole separations.

Kirova and Brazovskii 2004).

In this chapter we describe detailed investigations of the excited states of systems from small molecules to very long oligomers. These investigations use the DMRG method to solve the Pariser-Parr-Pople model, and thus are *essentially* assumption-free. The only assumptions are the relevance of parameters and the ability to use single-chain calculations to interpret experiments in the solid state. We find that both views concerning the relevance of electron-electron interactions have merits: some excited states can be viewed as weakly delocalized intraphenyl excitations, while other excited states (particularly the low energy states listed in Table 11.1) become important only in sufficiently large molecules. These are more conveniently interpreted from a semiconductor viewpoint.

11.2 Poly(*para*-phenylene)

It is evident from the chemical structure of the phenyl-based systems, shown in Fig. 1.2 and Fig. 11.3, that the phenyl ring is a key component of the structure. Since at the level of the π-electron approximation, phenyl and benzene rings are equivalent, a study of the electronic states of benzene provides useful insight to the electronic states of oligomers and polymers.

11.2.1 *Benzene*

We start by introducing the noninteracting description of benzene, although as we shall see, this description fails to explain the spectroscopic observations, indicating that electronic interactions are important in this molecule.

The noninteracting molecular orbitals of benzene have already been introduced in Chapter 3. These are the Bloch states,

$$|j\rangle = \frac{1}{\sqrt{6}} \sum_{n=1}^{6} c_n^\dagger \exp\left(-i\frac{\pi j n}{3}\right) |0\rangle, \tag{11.1}$$

where $c_n^\dagger$ creates an electron in the π-orbital on site n and the quantum number j satisfies, $j = 0, \pm 1, \pm 2, 3$. Equivalently, the molecular orbital wavefunctions are

$$\psi_j(\mathbf{r}) = \frac{1}{\sqrt{6}} \sum_{n=1}^{6} \phi_n(\mathbf{r}) \exp\left(-i\frac{\pi j n}{3}\right), \tag{11.2}$$

where $\phi_n(\mathbf{r})$ is a π-orbital on site n. The corresponding energies are

$$\epsilon_j = -2t \cos\left(\frac{\pi j}{3}\right). \tag{11.3}$$

In general, the molecular orbitals expressed by eqn (11.2) have complex amplitudes. Real amplitude molecular orbitals are trivially obtained by taking linear combinations of the degenerate pairs. In particular, we define the real amplitude molecular orbitals as

$$|1\rangle \equiv |j = 0\rangle;$$

$$|2\rangle \equiv \frac{1}{\sqrt{2}}\left(|j = 1\rangle + |j = -1\rangle\right);$$

$$|3\rangle \equiv \frac{1}{\sqrt{2}}\left(|j = 1\rangle - |j = -1\rangle\right);$$

$$|4\rangle \equiv \frac{1}{\sqrt{2}}\left(|j = 2\rangle + |j = -2\rangle\right);$$

$$|5\rangle \equiv \frac{1}{\sqrt{2}}\left(|j = 2\rangle - |j = -2\rangle\right);$$

$$|6\rangle \equiv |j = 3\rangle. \tag{11.4}$$

These real-amplitude molecular orbitals and their energies are illustrated in Fig. 11.4.

As discussed in Chapter 2, the π-electron models used in this book are invariant under the particle-hole transformation,

$$c_{i\sigma}^{\dagger} \rightarrow (-1)^{i} c_{i\bar{\sigma}}. \tag{11.5}$$

This implies that the molecular orbitals are related by a particle-hole transformation. In particular, numbering the sites as indicated in Fig. 11.4, it is readily shown that a molecular orbital $|\ell\rangle$ is transformed to its complement $|\bar{\ell}\rangle$ by the particle-hole transformation as follows,

$$|\ell\rangle \rightarrow -|\bar{\ell}\rangle. \tag{11.6}$$

The complementary pairs of orbitals connected in this way are,

$$|1\rangle \leftrightarrow -|6\rangle,$$

$$|2\rangle \leftrightarrow -|4\rangle,$$

and

$$|3\rangle \leftrightarrow -|5\rangle. \tag{11.7}$$

The ground state of benzene is determined by occupying the three lowest energy orbitals with two electrons each. Low-energy particle-hole excitations occur from the HOMOs (namely the states $|2\rangle$ and $|3\rangle$) to the LUMOs (namely the states $|4\rangle$ and $|5\rangle$). Thus, there are four degenerate excitations of energy $2t$. We denote a transition from $|\ell_h\rangle$ to $|\ell_e\rangle$ as $|\ell_e, \ell_h\rangle$. Then, from the nodal patterns of the molecular orbitals we see that excitations $|4_e, 2_h\rangle$ and $|5_e, 3_h\rangle$ are polarized along the z-direction, while the excitations $|5_e, 2_h\rangle$ and $|4_e, 3_h\rangle$ are polarized

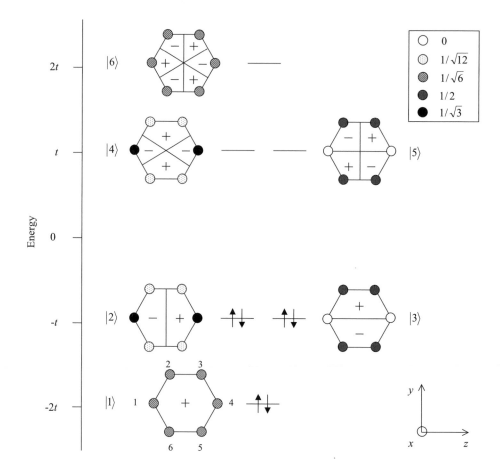

FIG. 11.4. The molecular orbitals of benzene and their electronic occupation in the
ground state. The shading indicates the atomic orbital amplitudes on each site. The
site labelling defines the particle-hole transformation rule, eqn (11.7).

along the y-direction. In fact, the *spatially allowed* excitations, $|1E_{1u}(z)\rangle$ and
$|1E_{1u}(y)\rangle$, are the linear combinations,

$$|1E_{1u}(z)\rangle = \frac{1}{\sqrt{2}}\left(|4_e, 2_h\rangle - |5_e, 3_h\rangle\right) \qquad (11.8)$$

and

$$|1E_{1u}(y)\rangle\rangle = \frac{1}{\sqrt{2}}\left(|5_e, 2_h\rangle + |4_e, 3_h\rangle\right), \qquad (11.9)$$

while the *spatially forbidden* excitations, $|1B_{1u}\rangle$ and $|1B_{2u}\rangle$, are the linear com-
binations,

Table 11.2 *The experimentally determined and theoretical predictions of the low-lying vertical excitations of benzene (in eV) (The experimental assignments are from (Bursill et al. 1998). Pariser-Parr-Pople calculations with $t = 2.539$ eV, and $U = 10.06$ eV (Bursill et al. 1998).)*

State	$|j|$	$\sigma(xy)$	$\sigma(xz)$	Experiment	Pariser-Parr-Pople model	CASPT2[a]
$1^1B_{2u}^+$	3	$+$	$-$	4.90	4.75	4.84
$1^1B_{1u}^-$	3	$-$	$+$	6.20	5.47	6.30
$1^1E_{1u}^-(z)$	1	$-$	$+$	6.94	6.99	7.03
$1^1E_{1u}^-(y)$	1	$+$	$-$	6.94	6.99	7.03
$1^3B_{1u}^+$	3	$-$	$+$	3.94	4.13	3.89
$1^3E_{1u}^+(z)$	1	$-$	$+$	4.76	4.76	4.49
$1^3E_{1u}^+(y)$	1	$+$	$-$	4.76	4.76	4.49
$1^3B_{2u}^-$	3	$+$	$-$	5.60	5.60	5.49

[a]Lorentzon *et al.* (1995). j is the angular momentum of the excited state, which is related to the Bloch momentum k, via $k = \pi j/3a$. The sign of $\sigma(xy)$ and $\sigma(xz)$ indicates the symmetry under a reflection through the xy or xz planes, respectively.

$$|1B_{1u}\rangle = \frac{1}{\sqrt{2}}\left(|4_e, 2_h\rangle + |5_e, 3_h\rangle\right) \tag{11.10}$$

and

$$|1B_{2u}\rangle = \frac{1}{\sqrt{2}}\left(|5_e, 2_h\rangle - |4_e, 3_h\rangle\right). \tag{11.11}$$

Next, we consider the particle-hole eigenvalues of these excited states. In analogy to the discussion of Section 2.9.2, using the rules in eqns (11.7), it is readily shown that the singlet/triplet $1E_{1u}(z)$, $1E_{1u}(y)$ and $1B_{1u}$ states have negative/positive particle-hole eigenvalues, while the singlet/triplet $1B_{2u}$ state has a positive/negative particle-hole eigenvalue. These assignments are shown in Table 11.2.

As usual, electronic interactions lift the degeneracies between the singlet and triplet states, and between states of different particle-hole symmetry. The experimentally determined transition energies shown in Table 11.2 indeed clearly deviate from the noninteracting prediction, indicating that electronic interactions play an important role in determining the character of the excited states. Rather than there being four degenerate singlet excitations, there are a pair of degenerate excitations and a further two excitations at different energies. The optically dominant singlet excitations are the spatially and particle-hole allowed $1^1E_{1u}^-(z)$ and $1^1E_{1u}^-(y)$ excitations at 6.94 eV. The spatially forbidden $1^1B_{1u}^-$ and $1^1B_{2u}^+$ excitations are weakly allowed because of vibronic coupling.

The $1^1B_{2u}^+$ excitation is particularly weak, because it is also forbidden by particle-hole symmetry selection rules. (These are weakly broken because of the lack of perfect particle-hole symmetry in conjugated systems.) We note that this state lies considerably lower in energy than the three higher states. This fact, and the positive particle-hole symmetry assignment, indicates that this state is

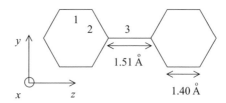

FIG. 11.5. The carbon sites in biphenyl. The bond labels define the hybridization integrals used in eqn (11.20). The torsion angle between adjacent phenylene rings is $\sim 30^0$ in solution.

highly correlated with 'covalent' character. Indeed, as described in Appendix G, the small excitation energy of this state is qualitatively predicted in the strong-coupling limit of the valence bond method where only covalent diagrams are retained. Rather than being a particle-hole excitation, the $1^1B_{2u}^+$ state is more correctly described as a linear superposition of the two equivalent Kekulé structures, as illustrated in Fig. G.1(b).

Also shown in Table 11.2 are the predictions from the Pariser-Parr-Pople model, where the parameters have been optimized to minimize the error on the excitations energies (Bursill *et al.* 1998). The optimized parameters are $U = 10.06$ eV and $t = 2.539$ eV. Fully *ab initio* CASPT2 predictions that are in good agreement with experiment are also shown (Lorentzon *et al.* 1995).

Finally, the low-lying triplet excitations of benzene are also shown in Table 11.2. We note that in contrast to the usual ordering of singlet and triplet states the $1^1B_{2u}^+$ state lies energetically below the $1^3B_{2u}^-$ state.

11.2.2 *Biphenyl*

We now turn to a discussion of the low-energy spectrum of biphenyl, again starting from the noninteracting limit. It is convenient to regard biphenyl as two benzene molecules (stripped of one hydrogen atom each) bonded together, as illustrated in Fig. 11.5. Since biphenyl possesses D_{2h} symmetry it is convenient to use the D_{2h} symmetry-adapted molecular orbitals of benzene (shown in Fig. 11.4) to construct its molecular orbitals.

We first note that the benzene molecular orbitals $|3\rangle$ and $|5\rangle$ have nodes in the wavefunction passing through the bridging atoms. Thus, these orbitals do not hybridize, and therefore become nonbonding biphenyl molecular orbitals. The remaining four molecular orbitals on each phenyl ring do hybridize, giving the bonding and antibonding molecular orbitals. These orbitals and their energies as shown in Fig. 11.6.

In the molecular orbital description the lowest-lying excitations of biphenyl are

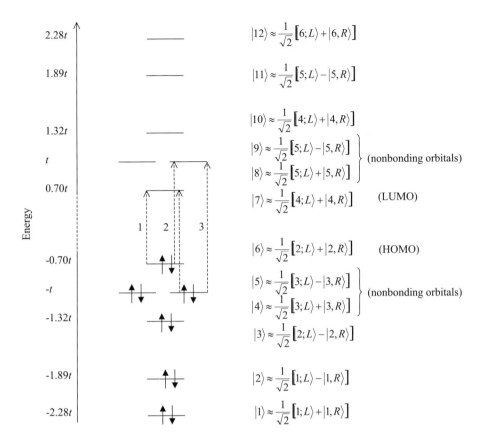

FIG. 11.6. Molecular orbitals of biphenyl represented by their predominant component left and right molecular orbitals of benzene (denoted by $|X; L\rangle$ and $|X; R\rangle$ and illustrated in Fig. 11.4). Also shown are the molecular orbital energies (using $t_s = t$), their electronic occupation in the ground state and the low-lying electronic transitions, labelled $1, 2,$ and 3.

1. The HOMO-LUMO excitations, denoted as $|7_e, 6_h\rangle\rangle$, at $1.4t$. This has B_{1u} symmetry and is polarized along the long-axis.[50]

2. The two degenerate excitations from the HOMO to the unoccupied nonbonding orbitals and the two degenerate excitations from the occupied nonbonding orbitals to the LUMO. These can be grouped as follows.

   (a) The excitation from the HOMO to the unoccupied *symmetric* nonbonding orbital, denoted as $|8_e, 6_h\rangle\rangle$, and the excitation from the occupied symmetric nonbonding orbital to the LUMO, denoted as

---

[50]To avoid confusion in this section, we identify biphenyl excitations with a $|\rangle\rangle$ symbol and benzene excitations with a $|\rangle$ symbol.

Table 11.3 *The experimentally determined and theoretical predictions of the low-lying vertical excitations of biphenyl (in eV) (The experimental assignments are from (Bursill et al. 1998). Pariser-Parr-Pople calculations with $t_p = 2.539$ eV, $t_s = 2.22$ eV, and $U = 10.06$ eV (Bursill et al. 1998). This table also serves to define the character table for D_{2h} symmetry group.)*

State	$\sigma(xy)$	$\sigma(xz)$	Experiment	Pariser-Parr-Pople model	CASPT2[a]
$1^1B_{3g}^+$	$-$	$-$	4.1 (0-0)	4.58	4.04
$1^1B_{2u}^+$	$+$	$-$	4.2 (0-0), 4.6	4.55	4.35
$1^1B_{1u}^-$	$-$	$+$	4.8	4.80	4.63
$1^1B_{3g}^-$	$-$	$-$	$-$	6.28	5.07
$1^1B_{2u}^-$	$+$	$-$	5.9	6.66	5.69
$2^1B_{1u}^-$	$-$	$+$	6.1	6.22	5.76
$2^1A_g^+$	$+$	$+$	ca. 6.0	6.30	5.85
$1^3B_{1u}^+$	$-$	$+$	ca. 3.5	3.63	3.10
$1^3B_{2u}^+$	$+$	$-$	3.9 (0-0)	4.56	4.14

[a]Lorentzon *et al.* (1995)

$|7_e, 4_h\rangle\rangle$, at $1.7t$. These have B_{2u} symmetry and are polarized along the short-axis. In addition, the linear combinations

$$\frac{1}{\sqrt{2}}(|8_e, 6_h\rangle\rangle \pm |7_e, 4_h\rangle\rangle) \qquad (11.12)$$

have $\mp$ particle-hole symmetry for the singlet states and $\pm$ particle-hole symmetry for the triplet states.

(b) The excitation from the HOMO to the unoccupied *antisymmetric* non-bonding orbital, denoted as $|9_e, 6_h\rangle\rangle$, and the excitation from the occupied antisymmetric nonbonding orbital to the LUMO, denoted as $|7_e, 5_h\rangle\rangle$, also at $1.7t$. These have B_{3g} symmetry.

3. The four degenerate excitations from the occupied to unoccupied nonbonding orbitals. In particular, we emphasize the excitation from the occupied to the unoccupied symmetric non-bonding orbitals, denoted as $|8_e, 4_h\rangle\rangle$, at $2t$, which has B_{1u} symmetry.

These excitations are shown in Fig. 11.6. In the noninteracting limit the biphenyl excitations $|7_e, 6_h\rangle\rangle$ and $|8_e, 4_h\rangle\rangle$ may be regarded as a *decoupling* of the *benzene* excitations $|4_e, 2_h\rangle$ and $|5_e, 3_h\rangle$.

As for benzene, electronic interactions in biphenyl significantly modify the noninteracting predictions. Table 11.3 lists the experimentally determined excitation energies. The degenerate pair of B_{2u} symmetry excitations are strongly split, with the $1^1B_{2u}^+$ state lying below the $1^1B_{1u}^-$ state. The biphenyl $1^1B_{2u}^+$ state is derived from its parent benzene $1^1B_{2u}^+$ state, and like its parent it is a strongly correlated 'covalent' state. Its positive particle-hole symmetry assignment in π-electron models means that it is only weakly dipole active in conjugated systems.

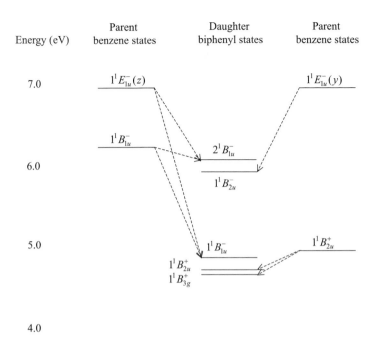

FIG. 11.7. Showing how the low-energy singlet excitations of benzene evolve to the significant low-energy singlet excitations of biphenyl. The molecular axes are defined in Fig. 11.5.

Similarly, the more dipole-active $1^1B_{2u}^-$ state is derived from the parent benzene $1^1E_{1u}^-(y)$ state.[51]

In contrast, the $1^1B_{1u}^-$ and $2^1B_{1u}^-$ states are derived from the mixing of the parent benzene $1^1B_{1u}^-$ and $1^1E_{1u}^-(z)$ states. As stated earlier, this mixing may be regarded as a decoupling of the intrabenzene excitations. Thus, a more useful way of understanding their origins is via the noninteracting picture, as follows. The $1^1B_{1u}^-$ state is related to the $|7_e, 6_h\rangle\rangle$ excitation. This becomes the $1^1B_{1u}^-$ exciton in polymers. The $2^1B_{1u}^-$ state is related to the $|8_e, 4_h\rangle\rangle$ excitation. This is the localized intraphenyl 'Frenkel' exciton.

The relation between the parent benzene states and the daughter biphenyl states is illustrated in Fig. 11.7.

The $1^1B_{3g}^+$ state, which is odd under both $\sigma(xy)$ and $\sigma(xz)$ reflection, and the $2^1A_g^+$ state, which is even under both $\sigma(xy)$ and $\sigma(xz)$ reflection, are also listed in Table 11.3. The $1^1B_{3g}^+$ state is the two-photon state associated with the

---

[51] At the Pariser-Parr-Pople model level of approximation the benzene $1^1B_{2u}^+$ and $1^1E_{1u}^-(y)$ states do not mix to form the biphenyl states, because they have opposite particle-hole symmetry.

transversely polarized $1^1 B_{2u}^+$ state. The $2^1 A_g^+$ state is the first excited symmetric even particle-hole symmetric state, which we argue becomes the $m^1 A_g$ state in polymers.

Table 11.3 also shows the predictions of the Pariser-Parr-Pople model (Bursill *et al.* 1998) and a CASPT2 calculation (Lorentzon *et al.* 1995). The additional parameter in the Pariser-Parr-Pople model, namely the bridging bond hybridization integral, t_s, is determined by fitting the predicted $1^1 B_{1u}^-$ transition energy to the experimental value. This gives $t_s = 2.22$ eV. With this fit the Pariser-Parr-Pople model does reasonably well at predicting the positions of the remaining states, except for the conspicuous failure of the $1^1 B_{2u}^-$, which is predicted to lie too high in energy and above the $2^1 B_{1u}^-$ state. The CASPT2 method is more successful at predicting the correct energetic ordering.

The four absorption peaks in light emitting polymers can be qualitatively understood by this investigation of the excited states of biphenyl. As we see in the next section, the $1^1 B_{2u}^+$ state is very weakly hybridized in oligomers and its energy remains virtually independent of chain length. In contrast, the $1^1 B_{1u}^-$ state strongly hybridizes, so its energy reduces with chain length to lie below the $1^1 B_{2u}^+$ energy at three or more phenyl rings. These excitations are responsible for peaks II and I, respectively. Likewise, peaks III and IV derive from the $1^1 B_{2u}^-$ and $2^1 B_{1u}^-$ states.

11.2.3   *Oligo and poly(para-phenylenes)*

Finally, we turn to describe the excited states of oligo and poly(*para*-phenylenes). As before, we briefly review the noninteracting description, before describing the affects of electronic interactions.

The analytical expression for the tight binding bands of poly(*para*-phenylene) is (Ambrosch-Draxl *et al.* 1995),

$$\epsilon_{1-4}(k) = \pm t_p \left(2\alpha^2 + \frac{1+\gamma^2}{2} \pm \beta \right)^{1/2}, \qquad (11.13)$$

where,

$$\beta^2 = \frac{1-\gamma^2}{2} + 4\alpha^2 \left(\frac{1+\gamma^2}{2} + \gamma \cos(kd) \right), \qquad (11.14)$$

$$\alpha = t_2/t_1, \qquad (11.15)$$

and

$$\gamma = t_3/t_1. \qquad (11.16)$$

The hybridization integrals are defined by the bond labels shown in Fig. 11.5. This spectrum is illustrated in Fig. 11.8. The low-energy zero-momentum excitations are also shown, in analogy to the low-energy excitations of biphenyl. These are

1. The $|d_{1e}^*, d_{1h}\rangle$ excitations, which have B_{1u} or A_g symmetry.

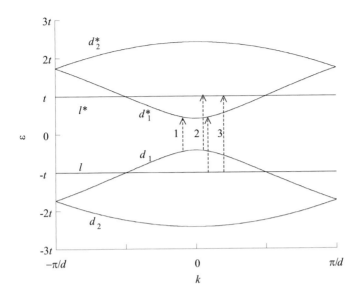

F<small>IG</small>. 11.8. The energy spectrum of poly(*para*-phenylene) using eqn (11.20) with $t_1 = t_2 = t_3 = t$. The low-lying electronic transitions are labelled 1, 2 and 3.

2. The degenerate excitations,

$$\frac{1}{\sqrt{2}} \left(|d^*_{1e}, l_h\rangle \pm |l^*_e, d_{1h}\rangle \right),\tag{11.17}$$

which have B_{2u} symmetry. For singlet excitations the symmetric combination has negative particle-hole symmetry, while the antisymmetric combination has positive particle-hole symmetry, and vice versa for triplet excitations.

3. The $|l^*_e, l_h\rangle$ excitation, which has B_{1u} symmetry.

Again, electronic interactions modify this picture, although for some excitations it is arguable that a noninteracting framework is a good starting point for the introduction of electronic interactions. In particular, the lowest energy excitations are excitonic, resulting from the attraction between the particle and hole in the $|d^*_{1e}, d_{1h}\rangle$ excitations. To see this, we review the Pariser-Parr-Pople model predictions obtained via the DMRG method (Bursill and Barford 2002). The parameters used in the calculation were, $t_1 = t_2 \equiv t_p = 2.539$ eV, $t_3 \equiv t_s = 2.22$ eV, and $U = 10.06$ eV.

Figure 11.9 shows the Pariser-Parr-Pople model predictions of the N-dependence of the transition energies of some key states. Also shown are experimental results for biphenyl, oligomers, and polymer thin films. To analyze these results it is useful to classify the excited states into a number of types.

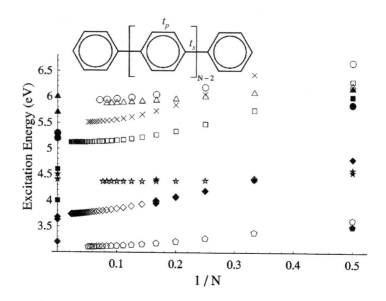

FIG. 11.9. The DMRG calculated transition energies of *para*-phenylene oligomers as a function of inverse chain length. Calculated from the Pariser-Parr-Pople model with unscreened parameters: $U = 10.06$ eV, $t_p = 2.539$ eV, $t_s = 2.22$ eV, and a dielectric constant, $\epsilon = 1$. $1^3 B_{1u}^+$ (pentagons), $1^1 B_{1u}^-$ (diamonds), $1^1 B_{2u}^+$ (stars), $2^1 A_g^+$ (squares), $1^3 A_g^-$ (crosses), the localized intraphenyl $^1 B_{1u}^-$ (Frenkel) state (triangles), and $1^1 B_{2u}^-$ (circles). The filled symbols are the experimental values for biphenyl, oligomers ($N = 3, \ldots, 6$ (Niko *et al.* 1999) and $N = 6$ (Zojer *et al.* 2000) and thin film polymers ($N = \infty$) (Lane *et al.* 1997; Niko *et al.* 1999). The inset shows the oligo-phenylene geometry with the bond integrals t_p and t_s. Reprinted with permission from R. J. Bursill and W. Barford, *Phys. Rev. B* **66**, 205112, 2002. Copyright 2002 by the American Physical Society.

11.2.3.1 $^1 B_{1u}^-$ *and* $^1 A_g^+$ *states associated with* $|d_{1e}^*, d_{1h}\rangle$ *excitations* These excited states are associated with the bound particle-hole excitations from the valence band to the conduction band. These excitations are essentially one-dimensional, because although the particle-hole wavefunction spreads over a phenyl ring, the centre-of-mass wavefunction propagates along the chain. We therefore expect that the Mott-Wannier exciton model described in Chapter 6 will apply to them.

As shown in Fig. 11.9, the calculated $1^1 B_{1u}^-$ energy indicates that this state strongly delocalizes. The DMRG results for the $1^1 B_{1u}^-$ energy in the $N = 3, \ldots, 6$ systems practically coincide with oligomer data. For large N the $1^1 B_{1u}^-$ energy approaches 3.73 eV in reasonable agreement with the experimental peak observed at 3.63–3.68 eV.

Table 11.4 *Calculated transition dipole moments connecting various* $^1A_g^+$ *and* $^1B_{1u}^-$ *states for the* $N = 8$ *system (For this oligomer the* $5^1B_{1u}^-$ *state is the Frenkel exciton, while the* $7^1B_{1u}^-$ *state is the 'n^1B_{1u}' state of the essential states model of nonlinear optical processes. The* $2^1A_g^+$ *state is always the 'm^1A_g' state in PPP. From (Bursill and Barford 2002).)*

j	$\langle 1^1A_g^+|\hat{\mu}|j^1B_{1u}^-\rangle$	$\langle j^1A_g^+|\hat{\mu}|1^1B_{1u}^-\rangle$	$\langle 2^1A_g^+|\hat{\mu}|j^1B_{1u}^-\rangle$
1	2.85	2.85	2.64
2	0.68	2.64	0.48
3	0.19	0.31	0.06
4	0.11	0.14	0.02
5	2.52	1.17	1.57
6	1.03	—	1.31
7	0.62	—	5.06
8	0.48	—	0.04

Electroabsorption studies place the $2^1A_g^+$ state at around 4.6 eV (Lane *et al.* 1997), approximately 0.5 eV below the extrapolated Pariser-Parr-Pople model result of 5.1 eV. This discrepancy may be explained by the characteristic red shifts generally observed for certain excited states when going from well isolated chains to polymers in the solid state (as described in Section 9.4). Typical estimates for this polarization or interchain screening shift are ~ 0.3 eV for the $1^1B_{1u}^-$ state and ~ 0.6 eV for the $2^1A_g^+$ state (Moore and Yaron 1998). These corrections resolve the theoretical and experimental predictions for the transition energy of the $2^1A_g^+$ state.[52]

Table 11.4 shows that the $2^1A_g^+$ state has a large transition dipole moment with the $1^1B_{1u}^-$ state, and unlike the case for polyenes, it is not predominantly a pair of bound magnons, but a particle-hole excitation. (It is usually labelled the m^1A_g state.) This particle-hole excitation is either an $n = 2$ Mott-Wannier exciton, or the edge of the unbound particle-hole continuum.

To investigate the position of a possible electron-hole continuum we consider the transition dipole moments between various $^1A_g^+$ states and the $1^1B_{1u}^-$ state, as well as between the $2^1A_g^+$ state and various $^1B_{1u}^-$ states. The $N = 8$ values, listed in Table 11.4, are representative of the general situation. We note that, in addition to the $1^1A_g^+$ and $2^1A_g^+$ states, another, higher lying state, which we denote as the $p^1A_g^+$ state, also has an appreciable transition dipole moment with the $1^1B_{1u}^-$ state. (For the $N = 8$ case $p = 5$.) There is also a pattern in the $\langle 2^1A_g^+|\hat{\mu}|j^1B_{1u}^-\rangle$ values. Namely, the $j = 1$ state has a strong transition dipole moment with the $2^1A_g^+$ state, as does the higher lying $^1B_{1u}^-$ absorption peak state (the localized intraphenyl exciton). In addition, there is another state,

---

[52]Qualitative modelling of solid state screening by using renormalized parameters, as discussed in Section 11.4, also corrects the transition energies (Chandross and Mazumdar 1997b, Bursill and Barford 2005).

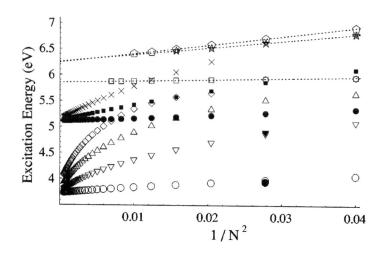

FIG. 11.10. The DMRG calculated transition energies in *para*-phenylene oligomers of a number of $^1A_g^+$ and $^1B_{1u}^-$ states as a function of $1/N^2$, where N is the number of repeat units. Calculated from the Pariser-Parr-Pople model with unscreened parameters: $U = 10.06$ eV, $t_p = 2.539$ eV, $t_d = 2.684$ eV, $t_s = 2.22$ eV, and $\epsilon = 1$. The low-lying $^1B_{1u}^-$ states are branches of the $n = 1$ family of Mott-Wannier excitons and the low-lying $^1A_g^+$ states are branches of the $n = 2$ family of Mott-Wannier excitons. (See also Fig. 6.5.) $1^1B_{1u}^-$ (large, open circles), $2^1B_{1u}^-$ (open, down triangles), $3^1B_{1u}^-$ (up triangles), $4^1B_{1u}^-$ (diamonds); $2^1A_g^+$ (small, solid circles), $3^1A_g^+$ (small, solid squares), $4^1A_g^+$ ($\times$). Also shown are the high lying localized intraphenyl $^1B_{1u}^-$ (Frenkel) excitation (open squares), $n^1B_{1u}^-$ (pentagons), and $p^1A_g^+$ (stars). The large, solid circle and solid down triangle show the position of the first and second long-axis polarized absorption peaks respectively for sexiphenyl ($N = 6$) (Zojer *et al.* 2000). The dotted lines are to guide the eye. Reprinted with permission from R. J. Bursill and W. Barford, *Phys. Rev. B* **66**, 205112, 2002. Copyright 2002 by the American Physical Society.

lying higher still, that has the largest transition dipole moment with the $2^1A_g^+$ state. We adopt the usual convention of denoting this state as the $n^1B_{1u}^-$ state. (In the $N = 8$ case $n = 7$.)

In order to further probe the nature of the various $^1A_g^+$ and $^1B_{1u}^-$ exciton and nonlinear optical states, we turn to an investigation of their large N behaviour. Fig. 11.10 shows a number of $^1A_g^+$ and $^1B_{1u}^-$ state transition energies as functions of $1/N^2$. Evidently the transition energies are linear in $1/N^2$ for large N. We see that there are a number of states in the $^1B_{1u}^-$ sector that converge to the same energy as the $1^1B_{1u}^-$ state in the long-chain limit. The ratio of their slopes is approximately $1 : 9 : 16$, etc. fitting the effective-particle exciton model for odd pseudomomentum quantum number j. This is the band of the $n = 1$

Mott-Wannier excitons described in Chapter 6. (Notice that the even pseudo-momentum quantum number j states, corresponding to the $^1A_g^-$ states, are not shown.)

Similarly, the $^1A_g^+$ sector has a number of odd pseudomomentum branches converging to the same energy. This is the band of the $n = 2$ Mott-Wannier excitons. (Again, the even pseudomomentum quantum number j states, in this case corresponding to the $^1B_u^+$ states, are not shown.) Notice the analogy of Fig. 11.10 to Fig. 6.5. A slice through energy for a particular oligomer size in Fig. 11.10 gives two bands of states associated with the principle quantum numbers, $n = 1$ and $n = 2$.

Above these states lie the $p^1A_g^+$ and the $n^1B_{1u}^-$ states, which converge to the same energy in the $N = \infty$ limit. The strong transition dipole moments from the $2^1A_g^+$ state to the $n^1B_{1u}^-$ state and from the $1^1B_{1u}^-$ state to the $p^1A_g^+$ state, and the close proximity in energy of the $p^1A_g^+$ and the $n^1B_{1u}^-$ states indicate that these states are close to the onset of the continuum of unbound particle-hole excitations. Lying below this continuum are the $n = 1$ and $n = 2$ Mott-Wannier excitons and the Frenkel exciton (as described below). The convergence of the $p^1A_g^+$ and the $n^1B_{1u}^-$ energies to ca. 6.25 eV as $N \to \infty$ would imply a very large binding energy (ca. 2.5 eV) for the $1^1B_{1u}^-$ exciton. However, band states are generally expected to be strongly affected by solid state screening (a red shift of 1.5 eV has been estimated for polyacetylene (Moore and Yaron 1998)). Taking such a shift into account would bring the $n^1B_{1u}^-$ energy and hence the exciton binding energy much closer to the results implied by electroabsorption experiments of ca. 1 eV. These corrections also imply a solid state binding energy for the $2^1A_g^+$ state of ca. 0.2 eV.

We conclude this section by remarking that we have exploited the particle-hole symmetry of the Pariser-Parr-Pople model to label the excited states with their particle-hole symmetries. In real conjugated systems, however, particle-hole symmetry is weakly broken (and more substantially broken with substituent side groups). Under these circumstances the state labels map onto those shown in Table 6.3. Thus, the $2^1A_g^+$ state becomes the m^1A_g state. This reflects the fact that even in a particle-hole symmetric model the $2^1A_g^+$ state is not necessarily the lowest even parity excited state, as in general there will be higher-lying pseudomomentum counterparts of the $1^1B_{1u}^-$ state with A_g^- symmetry that lie below the $2^1A_g^+$ state. This explains why even parity states with weak intensity are sometimes observed below the m^1A_g in light emitting polymers, as shown for example in Fig. 11.1(b).

11.2.3.2 $1^1B_{2u}^+$ state As already noted in Section 11.2.2, the particle-hole dipole-forbidden state $1^1B_{2u}^+$ lies below the dipole-active $1^1B_{1u}^-$ state in biphenyl. However, the $1^1B_{2u}^+$ state very weakly delocalizes, because the excited state wave-function has zero amplitudes on the bridging atoms, and delocalization therefore occurs via Coulomb-induced resonant exciton transfer (as described in Section 9.2). However, since the transition dipole moment with the ground state is very small for this state, resonant exciton transfer is not very effective. Thus, the energy of the $1^1B_{2u}^+$ state is almost independent of chain length, converging to 4.4 eV. This energy agrees well with the very weak second absorption peak at 4.4–4.5 eV (Lane *et al.* 1997) in polymer thin films. Adding weight to this interpretation is the observation of a weak but well defined 4.40 eV absorption peak in a highly textured film of sexiphenyl ($N = 6$), orientated perpendicular to the substrate (Zojer *et al.* 2000), as well as the weak, perpendicularly polarized absorption peak detected in orientated PFO film in the region 4.2–4.8 eV (Miller *et al.* 1999b).

The results of this and the last section indicate that the first (strong) and second (weak) absorption peaks in phenyl-based systems are the $1^1B_{1u}^-$ and $1^1B_{2u}^+$ states, respectively.[53]

11.2.3.3 $1^1B_{2u}^-$ **state** The third absorption peak is polarized normal to the long axis in PPP. In thin films it lies at 5.2–5.3 eV (Lane *et al.* 1997). A conspicuous failure of the Pariser-Parr-Pople model with the usual parametrization is its prediction for this state. The exact calculation for biphenyl places this state at 6.66 eV, whereas experimentally it is at ca. 5.85 (and below the $2^1B_{1u}^-$ state). As shown in Fig. 11.9 its calculated energy is 5.9 eV in the long chain limit, 0.6–0.7 eV higher than the experimental value.[54] As in biphenyl, this state derives from the $1^1E_{1u}^-(y)$ state of benzene. Like the $1^1B_{2u}^+$ state, it too delocalizes only via resonant exciton transfer, which is more effective for this state because of the larger transition dipole moment with the ground state.

11.2.3.4 *Frenkel exciton* The fourth absorption peak is polarized parallel to the long axis. In thin films it lies at 5.7–6.0 eV (Lane *et al.* 1997). This state is a highly localized intraphenyl (Frenkel) excitation which lies at 6.16 eV in biphenyl. Its N-dependence is plotted in Fig. 11.9.

11.2.3.5 *Other states* Fig. 11.9 also shows the lowest lying triplet states ($1^3B_{1u}^+$ and $1^3A_g^-$). We note that the $1^3A_g^-$ state lies around 0.4 eV higher than the $2^1A_g^+$ state. The close proximity of the $2^1A_g^+$ and $1^3A_g^-$ states is consistent with the

---

[53]Although the $1^1B_{2u}^+$ state is particle-hole dipole forbidden in the P-P-P model, it is (weakly) observable in biphenyl (and presumably larger systems) because particle-hole symmetry is actually broken in real systems. Another possible interpretation of the second absorption peak is that it is due to the $2^1B_{1u}^-$ state (see Chandross *et al.* 1997). That is, although the $2^1B_{1u}^-$ and $1^1B_{1u}^-$ states coincide in the $N = \infty$ limit (see Fig. 11.10), for systems of around $N = 8$ phenyl rings, the $2^1B_{1u}^-$ state has an appreciable transition dipole moment with the $1^1A_g^+$ state (see Table 11.4) and has an energy of around 4.4 eV. Although polydispersity would appear to rule out this scenario, this alternative interpretation is possible if we assume that the conjugation length distribution is sharply peaked around $N = 8$ rings in thin films (because, unlike the $1^1B_{2u}^+$ state, the $2^1B_{1u}^-$ state has strong N-dependence). Interestingly, beyond the first maximum centred at 3.95 eV, in addition to the short-axis polarized peak at 4.4 eV, the long-axis polarized absorption in Zojer *et al.* (2000) shows a peak at 4.91 eV that agrees well with the P-P-P $2^1B_{1u}^-$ result of 4.88 eV for $N = 6$ (see Fig. 11.10).

[54]These predictions are corrected by using a screened electron-electron interaction (Castleton and Barford 2002; Bursill and Barford 2005).

theory that in the weak-coupling Mott-Wannier exciton limit (as described in Chapter 6) these states are degenerate. However, because of the spin-density-wave contribution to its wavefunction the $2^1 A_g^+$ state lies lower in energy than the $1^3 A_g^-$ state.

11.3   Poly(*para*-phenylene vinylene)

Although poly(*para*-phenylene vinylene), or PPV, was the first phenyl-based polymer to exhibit electroluminescence its slightly more complicated chemical structure than PPP means that PPP is a more convenient model system to study theoretically. Nonetheless, the remarkable similarities in the optical spectroscopy of the two systems means that we should seek a common description of their excited states. Indeed, as we explain in this section, the theoretical description of the excited states of PPV, apart from overall energy differences, is very similar to PPP.

As shown in Fig. 1.2, PPV pocesses C_2 symmetry, and thus the states are classified as A_g or B_u. We begin this investigation by a study of stilbene, the smallest phenylene-vinylene oligomer.

11.3.1   *Stilbene*

Stilbene is represented in Fig. 11.12 by $N = 0$. Table 11.5 lists its experimental and calculated excitation energies. Since it is useful to relate these excitations to the corresponding excitations in biphenyl, the symmetry assignments of the states shown in brackets are the symmetries the stilbene states would have if stilbene had D_{2h} rather than C_2 symmetry. We make the same assignment of the origin of the four absorption peaks in PPV as for PPP. Namely, peak I originates from the $1^1 B_u^-$ (or $1^1 B_{1u}^-$) state, peak II originates from the $1^1 B_u^+$ (or $1^1 B_{2u}^+$) state, peak III originates from the $2^1 B_u^-$ (or $1^1 B_{2u}^-$) state, and peak IV originates from the $3^1 B_u^-$ (or $2^1 B_{1u}^-$) state.

11.3.2   *Oligo and poly(para-phenylene vinylenes)*

The band structure of poly(*para*-phenylene vinylene), derived using the method described in Appendix C, is shown in Fig. 11.11. Now there are eight bands arising from the eight π-orbitals per unit cell. The pair of nonbonding bands is a consequence of the D_{2h} symmetry of the Hückel model for PPV with short range transfer integrals. The low-lying excitations are precisely the same as those described in the previous section for PPP. As in PPP, the low-energy particle-hole excitations between the valence and conduction bands are responsible for the low energy delocalized excitons. Particle-hole excitations involving the nonbonding bands are responsible for the higher energy weakly delocalized excitons.

Figure 11.12 shows the DMRG calculated excitation energies of oligo(*para*-phenylene vinylenes) using the Pariser-Parr-Pople model with unscreened parameters (Bursill and Barford 2005).[55] As for oligo(*para*-phenylenes), the $1^1 B_u^-$

---

[55]In these calculations the geometry was straightened (i.e. the single-double bond angle was set to 180^0) so that the D_{2h} spatial symmetry could be used to target high-lying states. The

Table 11.5 *The experimentally determined and theoretical predictions of the low-lying vertical excitations of stilbene (in eV) (The symmetry assignments of the states shown in brackets are the symmetries the stilbene states would have if stilbene had D_{2h} rather than C_2 symmetry. The experimental assignments are from Castleton and Barford (1999). Pariser-Parr-Pople calculations with $t_p = 2.539$ eV, $t_d = 2.684$ eV, $t_s = 2.22$ eV, and $U = 10.06$ eV (Castleton and Barford 1999).)*

State	Experiment	Polarization	Pariser-Parr-Pople model
$1^1B_u^-$ ($\leftarrow 1^1B_{1u}^-$)	3.9, 4.2	Parallel	4.18
$1^1B_u^+$ ($\leftarrow 1^1B_{2u}^+$)	Higher than $1^1B_u^-$	Perpendicular	4.38
$2^1A_g^+$ ($\leftarrow 1^1B_{3g}^+$)	4.4, 4.5	—	4.39
$3^1A_g^+$ ($\leftarrow 2^1A_g^+$)	4.7, 5.0	—	5.12
$2^1B_u^-$ ($\leftarrow 1^1B_{2u}^-$)	5.4	Perpendicular	5.97
$3^1B_u^-$ ($\leftarrow 2^1B_{1u}^-$)	6.1	Parallel	5.80
$1^3B_u^+$ ($\leftarrow 1^3B_{1u}^+$)	2.3, 2.6	—	2.78

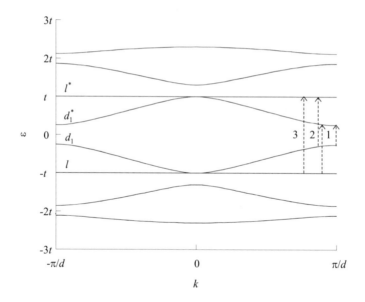

FIG. 11.11. The energy spectrum of poly(*para*-phenylene vinylene) with $t_1 = t_2 = t_3 = t$. The low-lying electronic transitions are labelled 1, 2 and 3.

and $2^1A_g^+$ states are the 1^1B_u and m^1A_g states, respectively, as indicated by the strong transition dipole moment between them. They are therefore the lowest

single and double bond lengths were shortened to preserve the overall molecular size. Thus $r_s = 1.283$ and $r_d = 1.194$.

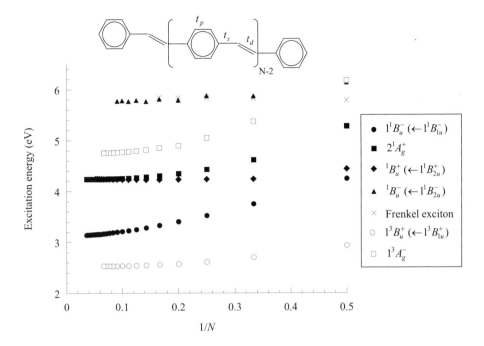

FIG. 11.12. The DMRG calculated transition energies of *para*-phenylene vinylene oligomers as a function of inverse chain length. Calculated from the
Pariser-Parr-Pople model with unscreened parameters: $U = 10.06$ eV, $t_p = 2.539$
eV, $t_d = 2.684$ eV, $t_s = 2.22$ eV, and dielectric constant, $\epsilon = 1$. The symmetry assignments of the states shown in brackets are the symmetries the PPV states would
have if PPV had D_{2h} rather than C_2 symmetry. The inset shows the oligo-phenylene
vinylene geometry with the bond integrals t_p, t_d and t_s.

pseudomomentum branches of $n = 1$ and $n = 2$ Mott-Wannnier excitons. The
higher-lying excitations, corresponding to the second, third and fourth absorption peaks are also shown. The second and third absorption peaks arise from
states that would have $^1B_{2u}^+$ and $^1B_{2u}^-$ symmetry if PPV had D_{2h} symmetry,
but instead have $^1B_u^+$ and $^1B_u^-$ symmetry. The fourth absorption peak is the
Frenkel exciton. Since these calculations were performed for a single chain using
the standard Pariser-Parr-Pople model parameters, our earlier discussion on the
importance of solvation effects also apply here. Namely, higher lying states, es-
pecially the m^1A_g and band states are expected to be strongly red-shifted in the
solid state. We discuss this point further in the next section.

11.4   Other theoretical approaches

Various theoretical approaches indicate two families of singlet excitons and two
families of triplet excitons below the conduction band threshold in PPV. Chan-

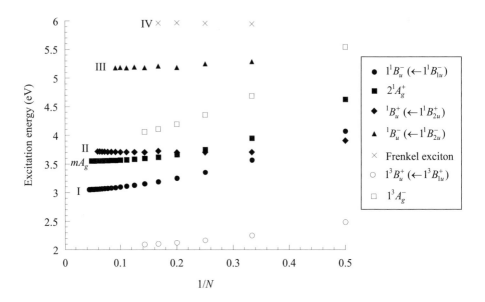

FIG. 11.13. The DMRG calculated transition energies of *para*-phenylene vinylene oligomers as a function of inverse chain length. Calculated from the
Pariser-Parr-Pople model with screened parameters: $U = 8$ eV, $t_p = 2.4$ eV, $t_d = 2.6$
eV, $t_s = 2.2$ eV, and $\epsilon = 2$. The excited states are identified with the spectroscopic
features shown in Fig. 11.1. (See also Fig. 11.14.) The symmetry assignments of the
states shown in brackets are the symmetries the PPV states would have if PPV had
D_{2h} rather than C_2 symmetry.

dross and Mazumdar (1997) solved the Pariser-Parr-Pople model at the single
configuration-interaction level using renormalized parameters. In particular, by
choosing $U = 8$ eV, $t_p = 2.4$ eV, $t_d = 2.6$ eV, $t_s = 2.2$ eV and a static dielectric function, $\epsilon = 2$ they were able to consistently fit the calculated single chain
absorption peaks of PPV to experiment. This renormalization can therefore be
regarded as a semiempirical modelling of the effects of solid state screening. For
an eight-unit oligomer they then calculate the 1^1B_u state at 2.7 eV, an m^1A_g
state at 3.3 eV and the n^1B_u state at 3.6 eV. The 1^1B_u and m^1A_g states are the
$n = 1$ and $n = 2$ excitons, while the n^1B_u state coincides with the charge-gap and
therefore indicates the onset of the particle-hole continuum. They also predict
the 1^3B_u state at 1.4 eV. These results are consistent with DMRG calculations
of the same model parameters shown in Fig. 11.13 (Bursill and Barford 2005).
Comparing this figure with Fig. 11.12 we see that the $1^1B_u^-$ and $2^1A_g^+$ states are
effectively solvated by ca. 0.1 eV and 0.6 eV, respectively. The excited states are
identified with the spectroscopic features shown in Fig. 11.1.

Beljonne *et al.* (1999) performed quantum chemical calculations using the
INDO Hamiltonian. They identified a number of important spectroscopic states,

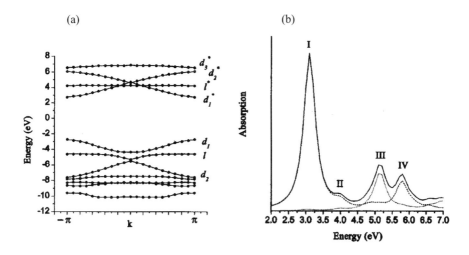

FIG. 11.14. (a) The π band structure and band labels for nonplanar di-hydroxy-PPV
calculated using the INDO Hamiltonian. (b) The associated absorption spectrum
using single configuration interactions. The solid line is the total absorption spectrum, the dotted line is absorption polarized perpendicular to the chain axis, and
the dashed line is absorption polarized parallel to the chain axis. The band compo-
sitions of the excited states are as follows: I, 97% $d_1 \rightarrow d_1^*$; II, 50% $d_1 \rightarrow \ell^*$, 35%
$\ell \rightarrow d_1^*$, 6% $d_1 \rightarrow d_1^*$; III, 44% $\ell \rightarrow d_1^*$, 34% $d_1 \rightarrow \ell^*$, 18% $d_1 \rightarrow d_1^*$; IV, 46% $\ell \rightarrow \ell^*$,
29% $d_1 \rightarrow \ell^*$, 11% $d_1 \rightarrow d_2^*$, 7% $\ell \rightarrow d_1^*$. Reprinted with permission from J. D.
Weibel and D. Yaron, *J. Chem. Phys.* **116**, 6846, 2002. Copyright 2002, American
Institute of Physics.

in particular the $1^1 B_u$ and $m^1 A_g$ states. An *ab initio* calculation by Rohlfing and
Louie (1999) on a PPV polymer predicts dipole allowed and forbidden singlet
excitons at 2.4 eV and 2.8 eV, respectively, with the quasi-particle gap at 3.3
eV. They also predict triplet excitons at 1.5 eV and 2.7 eV. The 2.4 eV and
2.8 eV singlet excitons are the $1^1 B_u$ and $m^1 A_g$ states, respectively, while the
1.5 eV and 2.7 eV triplet excitons are the $1^3 B_u$ and $m^3 A_g$ states, respectively.
The $m^1 A_g$ and $m^3 A_g$ states are nearly degenerate, as predicted by the Mott-
Wannier exciton theory for odd parity particle-hole wavefunctions. Using the
same technique with a screened electron-hole interaction van der Horst *et al.*
(2001) predict $1^1 B_u$ binding energies in ladder-PPP and PPV of 0.43 and 0.48
eV, respectively.

   The origin of the higher-lying peaks has also been investigated. Rohfling and
Louie (1999), and Weibel and Yaron (2002) predict that peak II in PPV arises
from an exciton caused predominately by the $(|d_{1e}^*, l_h\rangle - |l_e^*, d_{1h}\rangle)/\sqrt{2}$ particle-
hole excitation. This is essentially equivalent to the proposition that this peak

arises from the $1^1B_{2u}^+$ excitation of benzene (Rice and Garstein 1994). Weibel and Yaron (2002) have also investigated the effects of breaking particle-hole symmetry on the oscillator strength and polarization of peak II. Using the semiempirical INDO Hamiltonian on nonplanar di-hydroxy-PPV their calculations indicate that chemical substitution and mixing of the π and σ orbitals enhances the oscillator strength, as originally suggested by Gartstein *et al.* (1995). Moreover, as illustrated in Fig. 11.14, this peak becomes predominately polarized along the chain axis, in agreement with experiment (Miller *et al.* 1999a).

Most authors agree that the peak III can be assigned to an exciton caused by the $(|d_{1e}^*, l_h\rangle + |l_e^*, d_{1h}\rangle)/\sqrt{2}$ particle-hole excitation. This is essentially equivalent to the proposition that this peak arises from the $1^1E_{1u}^-(y)$ excitation of benzene. Similarly, peak IV is assigned to the Frenkel exciton caused by the $|l_e^*, l_h\rangle$ particle-hole excitation. These are the original assignments proposed by Rice and Garstein (1994).

Shukla *et al.* (2003) have investigated the high energy k^1A_g state by multi-reference configuration interactions, and argue that this state arises from double $d_1 \rightarrow l^*$ and $l \rightarrow d_1^*$ excitations.

11.5 The excited states of light emitting polymers

The experimental and theoretical studies of light emitting polymers described in this chapter suggest that the excited states can be understood as follows:

- Peak I corresponds to the low-energy dipole active $1^1B_u^-$ (or $1^1B_{1u}^-$) state. This is the lowest pseudomomentum branch of the family of $n = 1$ Mott-Wannier singlet excitons resulting from the Coulomb attraction between the particle-hole excitation from the valence (d_1) to the conduction (d_1^*) bands.

- Approximately 0.7 eV higher in energy is the m^1A_g state, identified by electroabsorption (Martin *et al.* 1999), two-photon absorption and photoinduced absorption (Frolov *et al.* 2002). The Pariser-Parr-Pople model calculations described in this chapter suggest that this state is the $2^1A_g^+$ state, which is the lowest pseudomomentum branch of the family of $n = 2$ Mott-Wannier excitons. This is sometimes labelled a charge-transfer exciton, because the particle-hole separation is greater than in the strongly bound 1^1B_u exciton. This assignment places a lower bound on the spectroscopically determined binding energy of the 1^1B_u exciton of 0.7 eV.

- Approximately 0.7 eV below the $1^1B_u^-$ exciton is the $1^3B_u^+$ triplet, indicating a large exchange energy characteristic of correlated states. This state is the lowest pseudomomentum branch of the family of $n = 1$ Mott-Wannier triplet excitons.

- Photo-induced absorption from the $1^3B_u^+$ triplet indicates another triplet, the $1^3A_g^-$ state, at approximately 1.4 eV higher in energy, and essentially degenerate with the $2^1A_g^+$ state. This triplet state is the lowest pseudomomentum branch of the family of $n = 2$ Mott-Wannier triplet excitons. As expected from Mott-Wannier exciton theory described in Chapter 6 the odd particle-hole parity singlet and triplet (charge-transfer) excitons are virtually degenerate.

- The n^1B_u state at 0.1 eV higher in energy than the m^1A_g state in PPV (Martin *et al.* 1999) indicates binding energies of ~ 0.8 eV and 0.1 eV for the $n = 1$ and $n = 2$ singlet excitons, respectively.

- Higher in energy are the excitations associated with peak II. The chain independent position of this peak, its transverse polarization in PPP and its small oscillator strength all indicate that it derives from the highly correlated $1^1B_{2u}^+$ state of benzene. Its optical strength arises from the breaking of particle-hole symmetry via substitution and $\pi - \sigma$ bond mixing.[56]

- The $k^1A_g^+$ state observed by two-photon absorption and photoinduced absorption slightly lower in energy than the $1^1B_{2u}^+$ state by Frolov *et al.* (2002) is long-lived and readily undergoes interchain charge separation. Although the origin of the $k^1A_g^+$ state is unclear, the experimental evidence suggests that it is the even-parity partner of the $1^1B_{2u}^+$ state (namely, the $1^1B_{3g}^+$ state in PPP, which has $^1A_g^+$ symmetry in PPV) (Shukla *et al.* 2003).

- Peak III is the $1^1B_{2u}^-$ state of PPP. This state derives from the $1^1E_{1u}^-(y)$ state of benzene.

- Finally, peak IV is the intraphenyl, or Frenkel, exciton.

11.6 Electron-lattice coupling

We now turn to a discussion of the effects of electron-lattice coupling on the electronic states of light emitting polymers.

Electron-lattice coupling has profound effects on the behaviour of conjugated polymers. It is responsible for the self-trapping of excited states. It also plays a vital role in determining the interconversion between excited states, and in energy and charge transfer processes. Predicting interconversion rates is important for understanding many electronic processes in conjugated polymers, for example, the determination of the singlet exciton yield in light emitting polymers, as described in Section 9.6

The phenyl-based conjugated polymers are extrinsically semiconducting as a consequence of the chemical structure determined by the σ bonds. Thus, with all bond lengths equal there is still a semiconducting band gap, as shown by Figs 11.8 and 11.11. However, coupling of the π-electrons to the lattice is still important as it causes the types of excited state structures described in Chapters 4 and 7 for extrinsically semiconducting polymers.

---

[56]However, the origin of this state is somewhat controversial. Chandross *et al.* (1997) argue that its predominately longitudinal polarization in PPV (Miller *et al.* 1999a) suggests that it is a higher momentum branch associated with the 1^1B_u state.

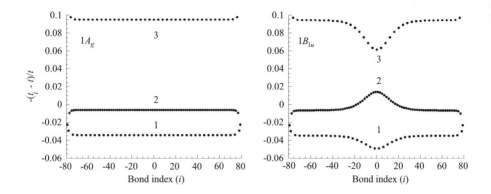

FIG. 11.15. The fractional change in transfer integrals of poly(*para*-phenylene) from
the uniform value, t, in the noninteracting limit. The electron-phonon parameter
used in the Peierls model (eqn (4.1) is $\lambda = 0.12$. The labels refer to the bonds shown
in Fig. 11.16. Only the upper rung of bonds are shown. Notice that the change in
transfer integrals is opposite to the change in bond lengths.

Electron-lattice coupling in light emitting polymers has been investigated by
a number of groups using a variety of methods, including multi-reference configuration interactions (Beljonne *et al.* 1995), density functional theory (Ambrosch-
Draxl *et al.* 1995), semiempirical Austin Model 1 (Zojer *et al.* 1999), time dependent Hartree-Fock theory (Tretiak *et al.* 2002), GWA-Bethe-Salpeter equation
(Artacho *et al.* 2004) and DMRG calculations of the Pariser-Parr-Pople-Peierls
model (Moore *et al.* 2005).

In this section we discuss the geometrical structures and relaxed energies of
the phenyl-based systems. We take the noninteracting and interacting electron
limits in turn.

11.6.1 *Noninteracting limit*

As discussed in Chapter 4, the noninteracting limit in the adiabatic approximation is described by the Peierls model (defined in Section 4.2). The ground and
excited state structures are easily obtained via the Hellmann-Feynman procedure, described in Section 4.4.

Figure 11.15 shows the fractional change in transfer integrals for the ground
state, δt_i, defined in eqn (4.5). The bonds are defined by Fig. 11.16. Since the
bonds are initially all of the same length, we see that the coupling of the π-
electrons to the lattice has caused an effective 'bond' alternation. The phenyl-
ring bonds shorten while the bridging bond lengthens. This is the benzenoid
structure, as the phenyl-ring bonds are roughly all of the same length.

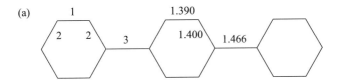

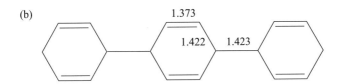

FIG. 11.16. (a) The bonds illustrated in Figs 11.15 and 11.19, and bond lengths in Å of the ground state determined in the interacting limit. (b) The quinoid structure of the $1^1B_u^-$ state. Bond lengths in the centre of the distortion in the interacting limit. (See Section 11.6.2 and Fig. 11.19)

To see this effective bond alternation we define the summed bond distortions as

$$\delta t_n = \sum_{i \in \text{phenyl ring}} \delta t_i; \text{ odd } n \tag{11.18}$$

and

$$\delta t_n = \delta t_{i=\text{ bridging bond}}; \text{ even } n. \tag{11.19}$$

Then we define the normalized, staggered and summed 'bond' alternation, δ_n, as

$$\delta_n = \frac{\delta t_n}{t}(-1)^n. \tag{11.20}$$

Figure 11.17 shows δ_n for the ground state. Under this mapping the phenyl ring is equivalent to a double bond (or dimer) and the bridging bond is a single bond. As for polyenes, this effective alternation increases the semiconducting band gap. Note that end-effects coupled to the constraint of an overall constant contour length causes the oscillations in δ_n: there are greater distortions in the phenyl rings at the end of the chain than in those in the middle of the chain. Thus, in the middle of the chain the summed distortion in bond lengths in a phenyl ring is not quite equal and opposite to the distortion of the bridging bonds.

Next consider the $1B_{1u}$ excited state structure, shown in Fig. 11.15. This is the quinoid structure, illustrated in Fig. 11.16 (b). In contrast to the ground state, there is now a significant variation in the bond lengths in the phenyl-ring: bonds labelled 1 shorten, while bonds labelled 2 lengthen. The bridging bond also shortens.

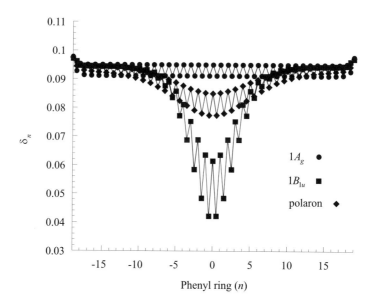

FIG. 11.17. The staggered, normalized and summed bond distortions of poly(*para*-phenylene) (as defined in eqn (11.20)) in the noninteracting limit.

At first sight the excited state lattice distortions of poly(*para*-phenylene) represented in Fig. 11.15 do not resemble those of a linear polyene. We therefore might enquire whether the geometrical defects (for example, solitons, polarons, etc.) and their associated mid-gap electronic states also exist in an analogous manner in poly(*para*-phenylene). To show that bond defects do exist in an analogous manner to linear polyenes we again consider the summed bond distortions, defined by eqns (11.18) - (11.20).

The $1B_{1u}$ state structure is illustrated in this way in Fig. 11.17. The relaxed $1B_{1u}$ state creates a 'polaronic' structure, whereby the average bond length in the phenyl ring increases while the bridging bond length decreases, but there is no reversal in sign of the bond distortions from the ground state. As described in Section 4.8, this polaronic structure of excited states occurs in extrinsically semi-conducting polymers where the ground state is nondegenerate: reversing the sign of the bond distortions gives a higher energy. A bond defect, or soliton, separates two regions of opposite bond distortions. Creating a soliton and antisoliton pair and moving them apart creates a region of reversed bonds. Thus, there is a linear confining potential between the soliton and antisoliton for large separations. As in linear polyenes, these bond defects are also associated with mid-gap states.

Associated with the two mid-gap single-particle states of the excited state are a bonding, ψ_i^+, and antibonding, ψ_i^-, molecular orbital (where i is a site index). As explained in Section 4.6, these molecular orbitals are analogous to the bonding and antibonding orbitals of molecular hydrogen. The molecular orbitals

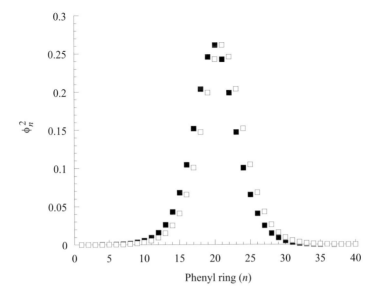

FIG. 11.18. The soliton (solid symbols) and antisoliton (open symbols) probability densities (defined by eqn (11.24)).

are constructed from localized Wannier functions, ϕ_i and $\bar{\phi}_i$, which represent the soliton and antisoliton respectively. In particular,

$$\psi_i^\pm = \frac{1}{\sqrt{2}}(\phi_i \pm \bar{\phi}_i), \tag{11.21}$$

or inverting

$$\phi_i = \frac{1}{\sqrt{2}}(\psi_i^+ + \psi_i^-) \tag{11.22}$$

and

$$\bar{\phi}_i = \frac{1}{\sqrt{2}}(\psi_i^+ - \psi_i^-). \tag{11.23}$$

In linear polyenes with degenerate ground states the soliton and antisoliton are widely separated. However, as described above, they are confined in extrinsic semiconductors. This confinement is illustrated for poly(*para*-phenylene) in Fig. 11.18, which shows the soliton and antisoliton probability density summed over each phenyl ring,

$$\phi_n^2 = \sum_{i \in \text{phenyl ring}} \phi_i^2. \tag{11.24}$$

We see that the soliton and antisoliton wavefunctions are centred on neighbouring phenyl rings in the middle of the chain.

Table 11.6 *The relaxation energies of the $1B_{1u}$ state and polaron for para-phenylene oligomers (in eV) calculated from the Peierls model (eqn (4.1)) (t = 2.514 eV and $\lambda = 0.12$)*

Number of phenyl rings	$1B_{1u}$ state	Polaron
4	0.23	0.06
8	0.15	0.04
20	0.08	0.02
40	0.06	0.01

This description of the molecular orbital defect states means that we can again apply the argument of Section 4.6 to describe the solitonic character of the excited states. In particular, the 1^1B_{1u} state is a linear superposition of spinless positively and negatively charged soliton-antisoliton pairs, while the 1^3B_{1u} state is a linear superposition of neutral spin-1/2 soliton-antisoliton pairs. These differences in the solitonic descriptions become important when the spin degeneracy is lifted by electronic interactions, and they help explain the quite different geometrical distortions of these two states in the interacting limit. We investigate these structures in the next section.

To aid in our understanding of exciton-polaron structures (to be described below) and relaxation energies, we also investigate charged (polaron) states in the noninteracting limit. Figure 11.17 shows the polaronic structure associated with a doped particle. Table 11.6 lists the relaxation energies of the $1B_{1u}$ state and polaron for different oligomer lengths. We note that the relaxation energy of the $1B_{1u}$ state is considerably greater than for the polaron, and that the relaxation energies reduce as the oligomer lengths increase.

In general, poly(*para*-phenylene) is not planar because of the steric repulsion of the hydrogen atoms. The torsional angle between adjacent phenyl rings for a single chain is estimated to be 27^0 (Ambrosch-Draxl *et al.* 1995). Packing in a crystalline environment planarizes the chain, and in this case the torsional angle is estimated to be 17^0. The quinoid structure of the excited state also planarizes the chain, because in this structure the bridging bond has more double bond character, and thus twisting the rings reduces the bond integral and hence increases the energy more than in the benzenoid structure. The torsional angle in the middle of the distortion is estimated to reduce to $\sim 8^0$ (Artacho *et al.* 2004).

11.6.2   *Interacting limit*

In Chapter 7 we described the combined effects of electron-lattice and electron-electron interactions on the electronic states of conjugated polymers. We discussed these effects for polymers with and without extrinsic semiconducting gaps. We saw that electron-electron interactions enhance the bond alternation in the ground state, and generally enhance the size of the lattice distortions for excited states. The enhancement is greater and the electron-lattice relaxation energy is larger for states with covalent character relative to states that are entirely ionic

Table 11.7 *The vertical and relaxation energies of para-phenylene oligomers calculated from the Pariser-Parr-Pople-Peierls model (eqn (7.1)) (t = 2.514 eV, U = 10.06 eV, and λ = 0.12)*

State	Vertical transition energy		Relaxation energy	
	$N = 4$	$N = 8$	$N = 4$	$N = 8$
$1^3B_{1u}^+$	3.29	3.17	0.58	0.41
$1^1B_{1u}^-$	4.21	3.96	0.17	0.06
$2^1A_g^+$	5.52	5.26	1.28	1.14

in character. Thus, the $1^3B_u^+$ and $2^1A_g^+$ states undergo a greater electron-lattice relaxation than the $1^1B_u^-$ state. These features also occur for the electronic states of light emitting polymers, as we now describe.

To investigate the combined effects of electron-lattice and electron-electron interactions we again employ the Pariser-Parr-Pople-Peierls model introduced in Section 7.2. Notice that the calculations described here do not describe free rotations of phenyl rings relative to one another. Thus, their applicability are to ladder poly(*para*-phenylene), where the stereochemistry causes the rings to have a planar geometry, or polymers in the solid state, where ring rotations are more restricted.

Table 11.7 lists the vertical and relaxed energies of the $1^3B_{1u}^+$, $1^1B_{1u}^-$ and $2^1A_g^+$ states for 4 and 8 ring *para*-phenylene oligomers. As in linear polyenes, the relaxation energy of the $1^1B_{1u}^-$ state is small, whereas the relaxation energy of the $1^3B_{1u}^+$ state is large. The experimentally determined relaxation energy of the $1^1B_u^-$ state in the related polymer poly(*para*-phenylene vinylene) has been reported as 0.07 eV by Liess *et al.* (1997). We may also deduce the relaxation energy in poly(*para*-phenylene) and ladder poly(*para*-phenylene) from Fig. 3 of Hertel *et al.* (2001) by noting that the ratio of the intensities of the $0-1$ to $0-0$ vibronic peaks in the absorption or emission spectra is S, the Huang-Rhys factor. The relaxation energy is then $\hbar\omega \times S$, where $\hbar\omega$ is the characteristic phonon frequency. Thus, using $S = 0.6$ for ladder poly(*para*-phenylene), $S = 1.2$ for poly(*para*-phenylene) and $\hbar\omega = 0.2$ eV gives relaxation energies of 0.12 eV and 0.24 eV for ladder poly(*para*-phenylene) and poly(*para*-phenylene), respectively. The larger relaxation energy for poly(*para*-phenylene) is expected, as the rings are free to rotate, and this result is consistent with a calculated value of 0.22 eV reported in Artacho *et al.* (2004). The relaxation energy of the $1^1B_{1u}^-$ state in the interacting limit is intermediate between the relaxation energy of the $1B_{1u}$ state and polaron in the noninteracting limit, as listed in Table 11.6. This illustrates the exciton-polaron nature of the $1^1B_{1u}^-$ state, as further discussed below.

The relaxation energy of the $2^1A_g^+$ state is also large, but not large enough to cause an energy level reversal of the $1^1B_{1u}^-$ and $2^1A_g^+$ states. The difference in relaxation energies between the $1^3B_{1u}^+$ and $1^1B_{1u}^-$ states increases the $0-0$ energy singlet-triplet exchange gap from the vertical gap of ~ 0.6 eV to ~ 0.9 eV, in good agreement with experiment (Köhler and Beljonne 2004). We also see

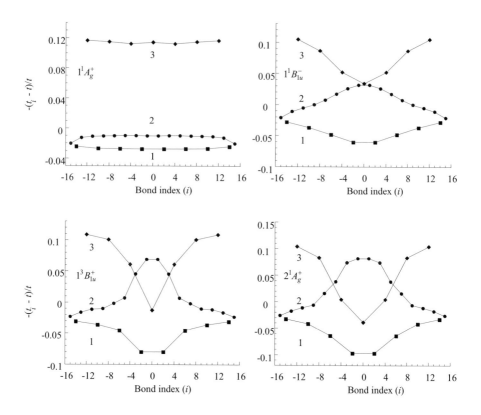

FIG. 11.19. The fractional change in transfer integrals of eight-ring *para*-phenylene oligomers from the uniform value, t, in the interacting limit. The parameters used in the Pariser-Parr-Pople-Peierls model (eqn (7.1)) are $U = 10.06$ eV, $t = 2.514$ eV, and $\lambda = 0.12$. The labels refer to the bonds shown in Fig. 11.16. Only the upper rung of bonds are shown.

that the relaxation energy reduces with chain size, consistent with an increased delocalization of the excitations and consequently a diminished effective electron-lattice coupling.

Next, we consider the associated geometrical structures. These are plotted in Fig. 11.19 for the normalized changes in transfer integrals and in Fig. 11.20 for the staggered, summed bond distortions. As predicted, the ground state alternation is enhanced in the interacting limit over the noninteracting limit by 8%. The bond lengths, calculated using $\delta r_i = -\delta t_i/\alpha$, are shown in Fig. 11.16.

The $1^1B_{1u}^-$ state is now an exciton-polaron. Its structure is qualitatively similar in both the noninteracting and interacting limits, as the soliton-antisoliton confinement due to linear confinement arising from the effective extrinsic bond alternation has a rather similar effect to electron-hole attraction. However, as already predicted, the $1^3B_{1u}^+$ state has a more pronounced distortion because it

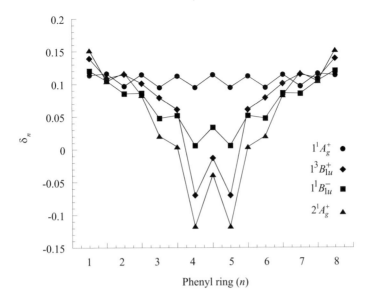

FIG. 11.20. The staggered, normalized and summed bond distortions of eight-ring *para*-phenylene oligomers (as defined in eqn (11.20)) calculated from the Pariser-Parr-Pople-Peierls model. $t = 2.514$ eV, $U = 10.06$ eV, and $\lambda = 0.12$.

has some covalent character. Indeed, there is a change of sign in the effective bond alternation. The middle bridging bond becomes a 'short' bond, while the adjacent phenyl-ring become 'long' bonds. Similarly, the $2^1 A_g^+$ state shows a significant structural distortion, with a change of sign of the bond alternation. The lattice distortions of the $1^1 B_{1u}^-$, $1^3 B_{1u}^+$, and $2^1 A_g^+$ states - as defined by the summed bond distortions of eqn (11.20) and shown in Fig. 11.20 - are qualitatively similar to those of linear polyenes with extrinsic dimerizarion, as described in Chapter 7.

The different relaxation energies and geometrical structures of the singlet and triplet B_{1u} states in the interacting limit is obviously related to the different kind of solitons comprising these states, as described in the previous section. In particular, the electronic interactions induce a strong coupling of the neutral soliton to the bond-order correlation, causing a significant distortion for the triplet state. In contrast, the charged solitons weakly couple to the bond-order correlation, and thus the singlet state is more weakly coupled to the lattice. Since the $2^1 A_g^+$ state has an admixture of charged and neutral solitons, it also couples more strongly to the lattice than the $1^1 B_{1u}^-$ state.

11.7 Concluding remarks

This chapter has described the spectroscopic and theoretical investigations of the excited states of light emitting polymers. Although the experiments are generally performed on polymers in thin film samples, the emphasis has been to describe the electronic properties of single polymer chains. The role of the environment has been discussed only in so far that it acts as a dielectric medium to screen the intramolecular excitations.

Intermolecular interactions, however, are crucial in determining the performance of light emitting and photovoltaic polymer devices. The effect of these interactions is also critically dependent on the local molecular structure. For example, as described in Section 9.2, if a pair of chains form a dimer in a parallel arrangement the lowest singlet excitation is necessarily dipole-forbidden. However, as described in Section 9.6, a parallel arrangement of chains is precisely the configuration required to optimize the singlet exciton yield. Interface states, for example, exciplexes, also play an important role in device performance. On a larger length scale, the overall polymer morphology - particularly for polymer blends - determines both charge and energy transport in polymer devices.

Understanding the precise role of structure on the performance of polymer electronic devices and other systems comprising conjugated polymers, for example light harvesting complexes, remains one of the outstanding challenges of the field.

APPENDIX A

DIRAC BRA-KET OPERATOR REPRESENTATION OF ONE-PARTICLE HAMILTONIANS

Throughout this book electronic models of conjugated polymers are developed within the number or second quantization representation. This representation is particularly powerful for treating many-body problems. However, as it is less familiar than a first quantization approach, this appendix explains the equivalence of the two approaches for single particle Hamiltonians. We take two examples: the fermion noninteracting (or Hückel) Hamiltonian and the exciton transfer model.

A.1 The Hückel Hamiltonian

The Hückel Hamiltonian, described in Chapter 3, is

$$H = - \sum_{n=1,\sigma}^{N} t_n \left(c_{n\sigma}^\dagger c_{n+1\sigma} + c_{n+1\sigma}^\dagger c_{n\sigma} \right), \tag{A.1}$$

where $t_n = t(1 + \delta_n)$.

As explained in Section 2.4, $c_{n\sigma}^\dagger$ creates an electron with spin σ in the spin-orbital, $\chi_n(\mathbf{r}, \sigma)$. We define the Dirac ket state as

$$|n, \sigma\rangle = c_{n\sigma}^\dagger |0\rangle, \tag{A.2}$$

where $|0\rangle$ is the vacuum state. The ket state $|n, \sigma\rangle$ is formally equivalent to the spin orbital $\chi_n(\mathbf{r}, \sigma)$.

The bra state, $\langle n, \sigma|$, is the conjugate to the ket state. The scalar product of a bra and ket is defined as

$$\langle m, \sigma | n, \sigma \rangle. \tag{A.3}$$

If the states form an orthonormal set then,

$$\langle m, \sigma | n, \sigma \rangle = \delta_{mn}. \tag{A.4}$$

For an orthonormal set we notice that the operator,

$$|m, \sigma\rangle\langle n, \sigma| \tag{A.5}$$

projects the state $|n, \sigma\rangle$ onto the state $|m, \sigma\rangle$:

$$|m, \sigma\rangle\langle n, \sigma | k, \sigma\rangle = |m, \sigma\rangle \delta_{nk}. \tag{A.6}$$

Thus, $|m, \sigma\rangle\langle n, \sigma|$ and $c_{m\sigma}^\dagger c_{n\sigma}$ are equivalent: both have the effect of transferring an electron from the spin-orbital $\chi_n(\mathbf{r}, \sigma)$ to the spin-orbital $\chi_m(\mathbf{r}, \sigma)$.

We can therefore express the Hückel Hamiltonian (eqn (A.1)) as,

$$H = - \sum_{n=1,\sigma}^{N} t_n \left(|n, \sigma\rangle\langle n+1, \sigma| + |n+1, \sigma\rangle\langle n, \sigma| \right). \tag{A.7}$$

As described in Section 3.3.1, for cyclic undimerized chains this Hamiltonian is diagonalized by the Bloch states,

$$|k, \sigma\rangle = \frac{1}{\sqrt{N}} \sum_{n} |n, \sigma\rangle \exp(ikna), \tag{A.8}$$

where N is the number of sites. To demonstrate this, consider

$$\begin{aligned}
H|k, \sigma\rangle &= H \frac{1}{\sqrt{N}} \sum_{n} |n, \sigma\rangle \exp(ikna) \\
&= -\frac{t}{\sqrt{N}} \sum_{n,\sigma} \left(|n, \sigma\rangle \exp(ik(n+1)a) + |n+1, \sigma\rangle \exp(ikna) \right),
\end{aligned} \tag{A.9}$$

where we have used the properties of the projection operator, eqn (A.5). Resumming the second term on the right-hand side, we have

$$\begin{aligned}
H|k, \sigma\rangle &= -\frac{t}{\sqrt{N}} \sum_{n,\sigma} \left(|n, \sigma\rangle \exp(ik(n+1)a) + |n, \sigma\rangle \exp(ik(n-1)a) \right), \\
&= -2t \cos(ka)|k, \sigma\rangle = \epsilon_k |k, \sigma\rangle. \tag{A.10}
\end{aligned}$$

A.2 The exciton transfer Hamiltonian

The exciton transfer Hamiltonian,

$$H = \sum_{mn} \left(J_{mn} E_m^\dagger E_n + J_{nm} E_n^\dagger E_m \right) + \Delta \sum_{m} E_m^\dagger E_m, \tag{A.11}$$

was introduced in Section 9.2.2. Here, $E_m^\dagger$ creates an exciton on the mth molecule. The ket state

$$|m\rangle = E_m^\dagger \prod_{n=1}^{N} |GS\rangle_n \tag{A.12}$$

represents an exciton localized on the mth molecule. (In analogy to eqn (A.2) we may regard $\prod_{n=1}^{N} |GS\rangle_n$ as the vacuum state.)

Equation (A.11) is equivalent to,

$$H = \sum_{mn} \left(J_{mn}|m\rangle\langle n| + J_{nm}|n\rangle\langle m| \right) + \Delta \sum_{m} |m\rangle\langle m|. \tag{A.13}$$

For a dimer it is easy to show that the symmetric and antisymmetric states

$$|+\rangle = \frac{1}{\sqrt{2}} \left(|1\rangle + |2\rangle \right) \tag{A.14}$$

and

$$|-\rangle = \frac{1}{\sqrt{2}} \left(|1\rangle - |2\rangle \right) \tag{A.15}$$

diagonalize H with energies $(\Delta + J)$ and $(\Delta - J)$, respectively (with $J_{mn} = J_{nm} = J$).

APPENDIX B

PARTICLE-HOLE SYMMETRY AND AVERAGE OCCUPATION NUMBER

Electron models with particle-hole symmetry satisfy the condition that

$$\langle \hat{N} - 1 \rangle = 0 \tag{B.1}$$

when the number of electrons equals the number of orbitals. The number operator, $\hat{N} = \sum_\sigma c_\sigma^\dagger c_\sigma$. Thus, the occupancy of each orbital satisfies,

$$\langle \hat{N} \rangle = 1. \tag{B.2}$$

To prove this, note that under a particle-hole transformation,

$$(\hat{N} - 1) \rightarrow -(\hat{N} - 1), \tag{B.3}$$

that is,

$$\hat{J}(\hat{N} - 1)\hat{J}^\dagger = -(\hat{N} - 1), \tag{B.4}$$

where $\hat{J}$ is the particle-hole operator, satisfying

$$\hat{J}^\dagger \hat{J} = 1. \tag{B.5}$$

Thus, the expectation value of $(\hat{N} - 1)$ is

$$\langle \hat{N} - 1 \rangle \equiv \langle \Psi | \hat{N} - 1 | \Psi \rangle$$
$$= \langle \Psi | \hat{J}^\dagger \hat{J} (\hat{N} - 1) \hat{J}^\dagger \hat{J} | \Psi \rangle, \tag{B.6}$$

using eqn (B.5).

Now, if an eigenstate of the Hamiltonian is also an eigenstate of $\hat{J}$, which is automatically the case for a state of definite charge when the number of charges equals the number of orbitals, then

$$\hat{J} | \Psi \rangle = J(\Psi) | \Psi \rangle, \tag{B.7}$$

where $J(\Psi)$ is the particle-hole eigenvalue, with values ± 1. Using the conjugate to eqn (B.7), namely,

$$\langle \Psi | \hat{J}^\dagger = J^*(\Psi) \langle \Psi |, \tag{B.8}$$

as well as eqn (B.7), eqn (B.6) becomes

$$\langle \hat{N} - 1 \rangle = \langle \Psi | J^*(\Psi) \hat{J} (\hat{N} - 1) \hat{J}^\dagger J(\Psi) | \Psi \rangle$$
$$= |J(\Psi)|^2 \langle \Psi | \hat{J} (\hat{N} - 1) \hat{J}^\dagger | \Psi \rangle. \tag{B.9}$$

Finally, using eqn (B.4) and $|J(\Psi)|^2 = 1$ we see that

$$\langle \hat{N} - 1 \rangle = -\langle \hat{N} - 1 \rangle, \tag{B.10}$$

thus proving eqn (B.1).

APPENDIX C

SINGLE-PARTICLE EIGENSOLUTIONS OF A PERIODIC POLYMER CHAIN

In this appendix we derive the eigenfunctions and eigenvalues of a noninteracting periodic polymer chain. We use a straightforward generalization of the method employed in Section 3.3.1 for the uniform cyclic chain.

We write the Hückel Hamiltonian for a polymer composed of a periodic sequence of monomers as,

$$H = \sum_m H_m + \sum_m \sum_n H_{mn}, \tag{C.1}$$

where H_m is the intramonomer one-particle Hamiltonian and H_{mn} is the intermonomer one-particle Hamiltonian. In particular,

$$
\begin{aligned}
H_m &= \sum_\sigma \underline{c}^\dagger_{m\sigma} \underline{\underline{H}}_m \underline{c}_{m\sigma} \\
&= \sum_\sigma \left(c^{1\dagger}_{m\sigma} \ c^{2\dagger}_{m\sigma} \ \cdots \right) \cdot
\begin{pmatrix}
t^{11}_m & t^{12}_m & \cdots \\
t^{21}_m & t^{22}_m & \cdots \\
\cdots & \cdots & \cdots
\end{pmatrix}
\cdot
\begin{pmatrix}
c^1_{m\sigma} \\
c^2_{m\sigma} \\
\cdots
\end{pmatrix},
\end{aligned}
\tag{C.2}
$$

where $c^{i\dagger}_{m\sigma}$ creates an electron on the ith site of the mth unit cell and t^{ij}_m is the intramonomer transfer integral between sites i and j.

Similarly,

$$
\begin{aligned}
H_{mn} &= \sum_\sigma \underline{c}^\dagger_{m\sigma} \underline{\underline{H}}_{mn} \underline{c}_{n\sigma} \\
&= \sum_\sigma \left(c^{1\dagger}_{m\sigma} \ c^{2\dagger}_{m\sigma} \ \cdots \right) \cdot
\begin{pmatrix}
t^{11}_{mn} & t^{12}_{mn} & \cdots \\
t^{21}_{mn} & t^{22}_{mn} & \cdots \\
\cdots & \cdots & \cdots
\end{pmatrix}
\cdot
\begin{pmatrix}
c^1_{n\sigma} \\
c^2_{2\sigma} \\
\cdots
\end{pmatrix},
\end{aligned}
\tag{C.3}
$$

where t^{ij}_{mn} is the intermonomer transfer integral between sites i and j on monomers m and n, respectively.

Exploiting the translational invariance of the polymer we introduce the Bloch transforms,

$$c^\dagger_{m\sigma} = \frac{1}{\sqrt{N}} \sum_k c^\dagger_{k\sigma} \exp(ikmd), \tag{C.4}$$

and

$$c_{m\sigma} = \frac{1}{\sqrt{N}} \sum_k c_{k\sigma} \exp(-ikmd), \tag{C.5}$$

where the Bloch wavevector, $k = 2\pi j/Nd$, d is the repeat distance, N is the number of unit cells and the quantum number j satisfies, $-N/2 \leq j \leq N/2$.

Substituting the Bloch tranforms into eqn (C.2) and eqn (C.3), and following the same procedure as in Section 3.3.1, we obtain the momentum space representation,

$$H = \sum_{k\sigma} c^\dagger_{k\sigma} \underline{\underline{H}}^m c_{k\sigma} + \sum_{k\sigma}\sum_{n} c^\dagger_{k\sigma} \underline{\underline{H}}^{mn} c_{k\sigma} \exp(-iknd)$$

$$= \sum_{k\sigma} c^\dagger_{k\sigma} \underline{\underline{H}}^0 c_{k\sigma}, \tag{C.6}$$

where

$$\underline{\underline{H}}^0 = \underline{\underline{H}}^m + \sum_{n} \underline{\underline{H}}^{mn} \exp(-iknd). \tag{C.7}$$

Now, if $\underline{S}$ is the unitary matrix that diagonalizes $\underline{\underline{H}}^0$ (and $\underline{S}^\dagger$ is its adjoint), we may write eqn (C.6) as,

$$H = \sum_{k\sigma} c^\dagger_{k\sigma} \underline{S}^\dagger \underline{S} \underline{\underline{H}}^0 \underline{S}^\dagger \underline{S} c_{k\sigma}$$

$$= \sum_{k\sigma} \tilde{c}^\dagger_{k\sigma} \underline{\tilde{H}}^0 \tilde{c}_{k\sigma}, \tag{C.8}$$

where

$$\underline{\tilde{H}}^0 = \underline{S}\underline{\underline{H}}^0\underline{S}^\dagger \tag{C.9}$$

is the diagonal representation of the Hamiltonian and

$$\tilde{c}^\dagger_{k\sigma} = c^\dagger_{k\sigma} \underline{S}^\dagger \tag{C.10}$$

are the diagonalized Bloch operators.

We now use this procedure to find the eigensolutions of the dimerized chain and poly(*para*-phenylene) as examples of the method.

C.1   Dimerized chain

With the sites of the unit cell labelled as shown in Fig. 3.3, eqn (C.6) becomes

$$H = \sum_{k\sigma} \begin{pmatrix} c^{1\dagger}_{k\sigma} & c^{2\dagger}_{k\sigma} \end{pmatrix} \cdot \begin{pmatrix} 0 & t_d + t_s \exp(-i2ka) \\ t_d + t_s \exp(i2ka) & 0 \end{pmatrix} \cdot \begin{pmatrix} c^1_{k\sigma} \\ c^2_{k\sigma} \end{pmatrix}, \tag{C.11}$$

where t_s and t_d are the transfer integrals for the single and double bonds, respectively.

By the similarity transformation (eqn (C.9)) H is diagonalized to

$$H = \sum_{k\sigma} \begin{pmatrix} c^{v\dagger}_{k\sigma} & c^{c\dagger}_{k\sigma} \end{pmatrix} \cdot \begin{pmatrix} \epsilon^v_k & 0 \\ 0 & \epsilon^c_k \end{pmatrix} \cdot \begin{pmatrix} c^v_{k\sigma} \\ c^c_{k\sigma} \end{pmatrix}, \tag{C.12}$$

where $c^{v\dagger}_{k\sigma}$ and $c^{c\dagger}_{k\sigma}$ are defined by eqns (3.19) and (3.20), respectively, and ϵ^v_k and ϵ^v_k by eqns (3.23) and (3.24), respectively.

C.2   poly(*para*-phenylene)

With the sites of the phenyl ring labelled as shown in Fig. 11.4, we have

$$
H =
\sum_{k\sigma} \begin{pmatrix} c_{k\sigma}^{1\dagger} & c_{k\sigma}^{2\dagger} & c_{k\sigma}^{3\dagger} & c_{k\sigma}^{4\dagger} & c_{k\sigma}^{5\dagger} & c_{k\sigma}^{6\dagger} \end{pmatrix} \cdot
\begin{pmatrix}
0 & t & 0 & t_s \exp(-ikd) & 0 & t \\
t & 0 & t & 0 & 0 & 0 \\
0 & t & 0 & t & 0 & 0 \\
t_s \exp(ikd) & 0 & t & 0 & t & 0 \\
0 & 0 & 0 & t & 0 & t \\
t & 0 & 0 & 0 & t & 0
\end{pmatrix}
\cdot
\begin{pmatrix}
c_{k\sigma}^{1} \\
c_{k\sigma}^{2} \\
c_{k\sigma}^{3} \\
c_{k\sigma}^{4} \\
c_{k\sigma}^{5} \\
c_{k\sigma}^{6}
\end{pmatrix},
$$

$$(C.13)$$

where t and t_s are the intraphenyl and bridging bond transfer integrals, respectively. The eigenvalues are given by eqn (11.20) and plotted in Fig. 11.8.

APPENDIX D

DERIVATION OF THE EFFECTIVE-PARTICLE SCHRÖDINGER EQUATION

The Schödinger equation for the effective-particle model of excitons was introduced in Chapter 6. In this appendix we derive that equation.

First we need to derive an exciton Hamiltonian in the weak-coupling limit. To do this it is necessary to recast the Pariser-Parr-Pople model (see Section 2.8.3),

$$H = -\sum_{i\sigma} t_i(c^\dagger_{i\sigma}c_{i+1\sigma} + c^\dagger_{i+1\sigma}c_{i\sigma}) \tag{D.1}$$

$$+U\sum_i\left(N_{i\uparrow} - \frac{1}{2}\right)\left(N_{i\downarrow} - \frac{1}{2}\right) + \frac{1}{2}\sum_{i\neq j}V_j(N_i - 1)(N_{i+j} - 1),$$

in a molecular orbital basis. The Ohno interaction, V_j, is defined in eqn (2.55).

As a simplification, we assume that the Wannier molecular orbitals for a linear, dimerized chain are localized on a particular dimer, that is,

$$c^{v\dagger}_{\ell\sigma} \approx \frac{1}{\sqrt{2}}(c^\dagger_{2\ell-1\sigma} + c^\dagger_{2\ell\sigma}) \tag{D.2}$$

for the valence band (bonding) molecular orbital, and

$$c^{c\dagger}_{\ell\sigma} \approx \frac{1}{\sqrt{2}}(c^\dagger_{2\ell-1\sigma} - c^\dagger_{2\ell\sigma}), \tag{D.3}$$

for the conduction band (antibonding) molecular orbital. ℓ is the unit cell index. The inverse relations are thus,

$$c^\dagger_{2\ell-1\sigma} \approx \frac{1}{\sqrt{2}}(c^{v\dagger}_{\ell\sigma} + c^{c\dagger}_{\ell\sigma}) \tag{D.4}$$

and

$$c^\dagger_{2\ell\sigma} \approx \frac{1}{\sqrt{2}}(c^{v\dagger}_{\ell\sigma} - c^{c\dagger}_{\ell\sigma}). \tag{D.5}$$

Substituting eqns (D.4) and (D.5) into the Pariser-Parr-Pople model, eqn (D.1), we obtain the molecular orbital exciton Hamiltonian,[57]

$$H_{\text{exciton}} = H_1 + H_2 + H_3. \tag{D.6}$$

We now describe the terms in H_{exciton} in turn.

---

[57]The molecular orbital Hamiltonian also contains terms that change the occupancy of the valence and conduction bands. However, as such terms do not connect basis states within the exciton subspace they are neglected.

H_1 is the single-electron Hamiltonian,

$$H_1 = -\sum_{\ell\gamma\sigma} \tilde{t}_{\gamma\gamma} \left(c_{\ell\sigma}^{\gamma\dagger} c_{\ell+1\sigma}^{\gamma} + c_{\ell+1\sigma}^{\gamma\dagger} c_{\ell\sigma}^{\gamma} \right) + \sum_{\ell\gamma} \epsilon_\gamma N_\ell^\gamma, \qquad (D.7)$$

where

$$N_{\ell\sigma}^\gamma = \sum_\sigma c_{\ell\sigma}^{\gamma\dagger} c_{\ell\sigma}^\gamma \qquad (D.8)$$

and the index γ indicates the conduction ($\gamma = c$) or valence ($\gamma = v$) band. The first term on the right-hand side represents the transfer of electrons between nearest neighbour dimers. The second term is the one-electron energy of the HOMO and LUMO states. $2\Delta = \epsilon_c - \epsilon_v$ is the HOMO-LUMO gap on a dimer.

Next, H_2 describes the Coulomb interactions,

$$H_2 = \tilde{U} \sum_{\ell\gamma} N_{\ell\uparrow}^\gamma N_{\ell\downarrow}^\gamma + \frac{\tilde{U}}{2} \sum_{\ell\gamma\neq\gamma'} N_\ell^\gamma N_\ell^{\gamma'} + \sum_{\ell\neq\ell'\gamma\gamma'} \tilde{V}_{\ell'} N_\ell^\gamma N_{\ell+\ell'}^{\gamma'}. \qquad (D.9)$$

The first two terms are the Coulomb interactions between electrons on the same dimer, while the third term is the Coulomb interaction between electrons on dimers ℓ' units apart.

Finally, H_3 describes the exchange interaction.

$$H_3 = -X \sum_{\ell\gamma\neq\gamma'} \left[\mathbf{S}_\ell^\gamma \cdot \mathbf{S}_\ell^{\gamma'} + \frac{1}{4}(N_\ell^\gamma - 1)(N_\ell^{\gamma'} - 1) \right], \qquad (D.10)$$

where

$$\mathbf{S}_\ell^\gamma = \sum_{\rho\rho'} c_{\ell\rho}^{\gamma\dagger} \sigma_{\rho\rho'} c_{\ell\rho'}^\gamma \qquad (D.11)$$

and σ are the Pauli spin matrices. This interaction arises from the usual mechanism that the electrons in a triplet state *on the same dimer* avoid each other, whereas electrons in a singlet state do not. Thus, the exchange energy is $2X = U - V_1$. (This is precisely the energy difference between the $^1B_u^-$ and $^3B_u^+$ states on a dimer, as shown in Section 5.5.)

In terms of the atomic orbital parameters the remaining molecular orbital parameters are

$$\tilde{t}_{vv} = -\tilde{t}_{cc} = \frac{t(1-\delta)}{2},$$

$$\epsilon_c = -\epsilon_v = t(1+\delta) \equiv \Delta,$$

$$\tilde{U} = \frac{U + V_1}{2},$$

and

$$\tilde{V}_\ell = \frac{V_{2\ell-1} + 2V_{2\ell} + V_{2\ell+1}}{4},$$

where we adopt the notation that when the atomic orbital parameters and molecular orbital parameters both use the same letter the molecular orbital parameters are distinguished by tildes.

The scalar product $\langle R+\frac{r}{2}, R-\frac{r}{2}|H_{\text{exciton}}|\Phi^{MW}\rangle$ gives the following difference equation for the exciton wavefunction, $\Phi(r, R)$,

$$-\tilde{t}\left(\Phi\left(r-d, R+\frac{d}{2}\right) + \Phi\left(r+d, R-\frac{d}{2}\right) + \Phi\left(r-d, R-\frac{d}{2}\right) + \Phi\left(r+d, R+\frac{d}{2}\right)\right)$$
$$+ \left(2X\delta_{r0}\delta_M - \tilde{V}(r)\right)\Phi(r, R) = \left(E - \tilde{U} - 2\Delta + X\right)\Phi(r, R).$$

(D.1

$\delta_M = 1$ for singlet excitons and $\delta_M = 0$ for triplet excitons. $\delta_{r0} = 1$ when $r = 0$. d is the contour length between repeat units (e.g. $2a$ for a dimerized chain, where a is the lattice spacing), and

$$\tilde{t} = \frac{t(1-\delta)}{2}.$$

(D.13)

To derive an effective-particle model we separate the centre-of-mass and relative coordinates. For periodic boundary conditions we assume that

$$\Phi_{nK}(r, R) = \frac{1}{\sqrt{N_u}}\exp(iKR)\psi_n(r),$$

(D.14)

where K is the centre-of-mass momentum: $-\pi/d \leq K \leq \pi/d$.

For open boundary conditions we assume that

$$\Phi_{nj}(r, R) = \sqrt{\frac{2}{N_u + 1}}\sin(\beta_j R)\psi_n(r),$$

(D.15)

where β_j is the centre-of-mass pseudo-momentum:

$$\beta_j = \frac{j\pi}{(N_u + 1)d},$$

(D.16)

and $j = 1, 2, \cdots, N_u$.

Substituting eqn (D.14) into eqn (D.12), we obtain the following Schrödinger difference equation for the relative wavefunction, $\psi_n(r)$:

$$-2\tilde{t}\cos\left(\frac{Kd}{2}\right)(\psi_n(r+d) + \psi_n(r-d)) + \left(2X\delta_{r0}\delta_M - \tilde{V}(r)\right)\psi_n(r)$$
$$= \left(E - \tilde{U} - 2\Delta + X\right)\psi_n(r).$$

(D.17)

$\psi_n(r)$ is the relative wavefunction for the electron-hole pair in the localized molecular orbitals r/d molecular repeat units apart. A similar equation is obtained using eqn (D.15) with K replaced by β_j.

As $r \to \infty$

$$\tilde{V}(r) \to V(\epsilon_{\text{eff}}r),$$

(D.18)

where $V(r)$ is the Ohno potential, defined in eqn (2.55) and ϵ_{eff} is the effective dielectric constant arising from the polymer geometry. This scale factor arises because the electron-hole separation, or the relative coordinate, r, is measured as a contour length along the polymer chain (so, r/d is the number of repeat units between the electron and hole). However, the Coulomb interaction is determined by the geometrical separation between the electron and hole. The scaling between these length scales is determined by ϵ_{eff}. (e.g. in the *trans*-polyacetylene structure if the contour-distance between the electron and hole is r the geometrical separation is $r/\sqrt{3}$ and thus, $\epsilon_{\text{eff}} = 1/\sqrt{3}$.)

Equation (D.17) is solved in the following appendix in the effective-mass limit.

APPENDIX E

HYDROGENIC SOLUTIONS OF THE EFFECTIVE-PARTICLE EXCITON MODELS

In this appendix we examine the properties of the effective-particle exciton models derived in Appendix D and described in Chapter 6 in the continuum or effective-mass limit.

E.1 The weak-coupling limit

In making the connection to the continuum limit it is convenient to set $\tilde{t} = \hbar^2/2Md^2$, so $2\tilde{t} = \hbar^2/2\mu d^2$, where the reduced mass, $\mu = M/2$, and M is the effective mass. d is the contour length between repeat units (e.g. $2a$ for a dimerized chain).

Then, scaling lengths by the effective Bohr radius,

$$a_0(K) = \frac{4\pi\epsilon_0\epsilon\epsilon_{\text{eff}}\hbar^2\cos(Kd/2)}{\mu e^2}$$

(E.1)

and the energy by the effective Rydberg,

$$E_I(K) = \frac{\mu e^4}{2\hbar^2\cos(Kd/2)(4\pi\epsilon_0\epsilon\epsilon_{\text{eff}})^2},$$

(E.2)

eqn (D.17) becomes,

$$-\frac{1}{a'^2}(\psi_n(r'+a') + \psi_n(r'-a')) - \frac{2\psi_n(r')}{a'(1+r'^2)^{1/2}} = (E'_n - \tilde{U}' - 2\Delta')\psi_n(r'),$$

(E.3)

where $r' = r/a_0(K)$, $a' = d/a_0(K)$, $E'_n = E_n/E_I(K)$, $\tilde{U}' = \tilde{U}/E_I(K)$, and $\Delta' = \Delta/E_I(K)$. We have used the Ohno function (eqn (2.55)) for the Coulomb interaction, which remains finite as $r' \to 0$, and we set $X = 0$, as we are uninterested in the details of the exchange interaction. ϵ is the actual dielectric function, while ϵ_{eff} is the effective dielectric constant arising from the polymer geometry, as explained in Appendix D.

In the continuum limit ($a'(K) \to 0$) eqn (E.3) is identical to the effective one-dimensional equation for the radial part of the three-dimensional hydrogen atom wavefunction, $u(r) = r\phi(r)$, for the case of zero angular momentum, where $\phi(r)$ is the radial wavefunction (see Cohen-Tannoudji et al. (1977, p. 792)). This equation was studied in detail by Loudon (1959). It is useful to treat the even and odd parity solutions separately.

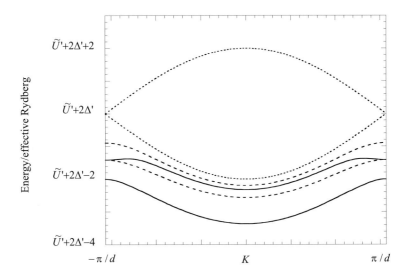

FIG. E.1. The dispersion curves of the four lowest bound states (solid and dashed) for a regularized Coulomb potential with $a'(K = 0) = 1$. Even (odd n) parity states (solid curves) and odd (even n) parity states (dashed curves). The particle-hole continuum is bounded by the dotted curves. The energies are in units of E_I.

E.1.1 *Odd parity, even n solutions*

The odd parity states have the same boundary conditions as $u(r)$, namely $u(0) = 0$ and $u(r \to \infty) \to 0$. They are formed by matching $u(r)$ with $-u(-r)$ at the origin. Thus, for even n the binding energies are

$$E_n(K) = \frac{E_I(K)}{(n/2)^2} \tag{E.4}$$

and the corresponding wavefunctions are

$$\psi_n(r, K) = (Nr/a'(K)) \exp\left(-2r/na'(K)\right) L_{n/2}\left(4r/na'(K)\right), \tag{E.5}$$

where L_m is the mth order Laguerre polynomial and N is a normalization constant.

Notice that as a result of the K dependency of $E_I(K)$ the binding energies for a given n are larger for the higher centre-of-mass momentum states. Similarly, the characteristic length, $a_0(K)$, decreases for higher momentum states, resulting in a smaller particle-hole separation.

The particle-hole continuum is trivially found by setting $\tilde{V}(r = 0)$ in eqn (D.17). The onset of the particle-hole continuum is at $\tilde{U} + 2\Delta - 2E_I(K)/a'(K)^2$, so the exciton energies relative to the ground state are

$$E_n^{\text{ex}}(K) = \tilde{U} + 2\Delta - \frac{2E_I(K)}{a'(K)^2} - E_n(K). \tag{E.6}$$

As $K \to 0$ we find,

$$E_n^{\text{ex}}(K) = E_0 - \frac{E_I(K=0)}{(n/2)^2} \left(1 + \frac{K^2 d^2}{8}\right) + \frac{\hbar^2 K^2}{2(2M)}, \tag{E.7}$$

where $E_0 = \tilde{U} + 2\Delta - 2E_I(K=0)/a'(K=0)^2$, and the last term is the kinetic energy of the effective particle of mass $2M$.

The average particle-hole separation, r_0, may be found from the expectation value of r,

$$r_0 = 2\langle r \rangle_n = 3(n/2)^2 a_0(K), \tag{E.8}$$

which increases rapidly with the principal quantum number, n, but decreases for higher momentum states of the same n.

E.1.2   *Even parity, odd n solutions*

The even parity wavefunctions do not satisfy the same boundary conditions as hydrogen atom wavefunction, $u(r)$, at $r = 0$, so there are no semianalytical results for these states. However, it can be shown that the lowest even parity state is strongly bound, with a binding energy scaling as $2/a'(K)$, while the energies of the remaining even parity states are bounded by a higher and lower odd parity state.

E.1.3   *Numerical results*

For arbitrary $a'(K)$ it is necessary to solve eqn (E.3) numerically. Figure E.1 shows the dispersion of the four lowest bound states and the particle-hole continuum for the value $a'(K=0) = 1$. Figure E.2 shows the binding energy of the four lowest states at $K = 0$ as a function of a'. As a' decreases the binding energies approach the Rydberg series, except for the energy of the first even parity state ($n = 1$), which diverges. We see that the $n = 1$ bound state is split-off from the remaining states, whose energies are scaled by the Rydberg energy.

Typical values for conjugated polymers, with $t = 2.5$ eV, are (i) $\delta = 0.1$ gives $E_I = 3.90$ eV, and $a' = 1.31$; (ii) $\delta = 0.2$ gives $E_I = 4.40$ eV and $a' = 1.48$. The resulting binding energies from Fig. E.1 agree very well with those of Fig. 5.1(a).

The results discussed here apply to an infinite, periodic chain. Identical results are obtained for infinite, linear chains, except that the centre-of-mass momentum, K, is replaced by the pseudo-momentum $\beta_j = j\pi/(N_u + 1)d$. Replacing K by β_j, eqn (E.7) indicates that the exciton energies scale as $(j/N_u)^2$ in the large N_u limit.

E.2   The strong-coupling limit

The analysis for this limit is very similar to that of the weak-coupling limit, except that now the hardcore repulsion imposes the boundary condition that $\psi(0) = 0$ on all the solutions. Thus, degenerate pairs of even and odd solutions are found by matching $\psi(r)$ with $\pm\psi(-r)$ as $r \to 0$. So, setting the molecular-orbital

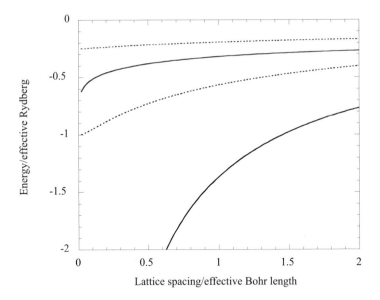

FIG. E.2. The zero-momentum exciton binding energies in units of E_I in the weak-coupling limit for a regularized Coulomb potential versus a/a_I. Even parity (odd n) states (solid curves) and odd (even n) parity states (dashed curves). The energies of the odd parity solutions approach the Rydberg series as $a/a_I \to 0$, while the energy of the $n = 1$ solution diverges.

parameters, denoted by the tilde, to the atomic-orbital parameters, $\Delta = 0$, the repeat distance, d, to a, and replacing $n/2$ by n on the right-hand sides of eqns (E.4) - (E.8) the strong-coupling results trivially follow:

$$a_0(K) = \frac{\hbar^2 \cos(Ka/2)\epsilon_{\text{eff}}}{\mu e^2}, \tag{E.9}$$

$$E_I(K) = \frac{\mu e^4}{2\hbar^2 \cos(Ka/2)\epsilon_{\text{eff}}^2}, \tag{E.10}$$

$$E_n(K) = \frac{E_I(K)}{n^2}, \tag{E.11}$$

and

$$E_n^{\text{ex}}(K) = U - \frac{2E_I(K)}{a'(K)^2} - E_n(K), \tag{E.12}$$

where $a'(K) = a/a_0(K)$.

APPENDIX F

EVALUATION OF THE ELECTRONIC TRANSITION DIPOLE MOMENTS

In this appendix we use the effective-particle exciton models introduced in Chapter 6 to calculate transition dipole moments. These results are summarized in Chapter 8.

F.1 The weak-coupling limit

In the weak-coupling limit a general excited state is of the form,

$$|p\rangle = \sum_{nj} \alpha^p_{nj} |\Phi^{MW}_{nj}\rangle, \tag{F.1}$$

where

$$|\Phi^{MW}_{nj}\rangle = \sum_{r,R} \psi_n(r)\Psi_j(R)|R + 2/2, R - r/2\rangle \tag{F.2}$$

and

$$|R + r/2, R - r/2\rangle = \frac{1}{\sqrt{2}} \left(c^{c\dagger}_{R+r/2,\uparrow} c^{v}_{R-r/2,\uparrow} \pm c^{c\dagger}_{R+r/2,\downarrow} c^{v}_{R-r/2,\downarrow} \right) |GS\rangle. \tag{F.3}$$

$|GS\rangle$ is the ground state, defined by eqn (6.9). $\psi_n(r)$ is the dimensionless 'hydrogenic' wavefunction for the particle-hole pair, where r is the relative coordinate.

$$\Psi_j(R) = \sqrt{\frac{2}{N+1}} \sin(\beta_j R) \tag{F.4}$$

is the centre-of-mass envelope wavefunction, where R is the centre-of-mass coordinate. $\beta_j = \pi j/(N_u + 1)$, where N_u is the number of unit cells. For simplicity, we now consider 'pure states', that is states with just one component of $|\Phi^{MW}_{nj}\rangle$ in $|p\rangle$.

The electronic dipole operator in one-dimension is

$$\hat{\mu}_e = e\hat{x} = e \sum_{\text{unit cells, } \ell} \left(x_{\ell 1}(\hat{N}_{\ell 1} - 1) + x_{\ell 2}(\hat{N}_{\ell 2} - 1) \right), \tag{F.5}$$

where the subscripts 1 and 2 refer to the left and right sites of the unit cell, respectively, as shown in Fig. 3.3.

Using the relations eqn (3.26) the density operators become

$$\hat{N}_{\ell 1} = \frac{1}{2}\left(\hat{N}_{\ell}^{v} + \hat{N}_{\ell}^{c} + \sum_{\sigma}(c_{\ell\sigma}^{v\dagger}c_{\ell\sigma}^{c} + c_{\ell\sigma}^{c\dagger}c_{\ell\sigma}^{v})\right) \qquad (F.6)$$

and

$$\hat{N}_{\ell 2} = \frac{1}{2}\left(\hat{N}_{\ell}^{v} + \hat{N}_{\ell}^{c} - \sum_{\sigma}(c_{\ell\sigma}^{v\dagger}c_{\ell\sigma}^{c} + c_{\ell\sigma}^{c\dagger}c_{\ell\sigma}^{v})\right). \qquad (F.7)$$

We now use these equations to calculate the matrix elements for the transitions.

F.1.1   *Transitions between the ground state and an excited state*

The only term in the dipole operator that connects the ground state to excited states is $c_{\ell\sigma}^{v\dagger}c_{\ell\sigma}^{c}$. So,

$$\hat{x} = \frac{1}{2}\sum_{\ell\sigma}\left(x_{\ell 1}c_{\ell\sigma}^{v\dagger}c_{\ell\sigma}^{c} - x_{\ell 2}c_{\ell\sigma}^{v\dagger}c_{\ell\sigma}^{c}\right),$$

$$= -\frac{a}{2}\sum_{\ell\sigma}c_{\ell\sigma}^{v\dagger}c_{\ell\sigma}^{c}, \qquad (F.8)$$

where a is the lattice parameter. The dipole operator can only connect a basis state $|R+r/2, R-r/2\rangle$ to $|GS\rangle$ if $r = 0$.

Thus,

$$\langle \text{GS}|\hat{x}|p\rangle = -\frac{a}{\sqrt{N_u + 1}}\sum_{R}\sin(\beta_j R)\psi_n(0). \qquad (F.9)$$

Summing over R we have,

$$\langle \text{GS}|\hat{x}|p\rangle = 0 \qquad (F.10)$$

for even j and

$$\langle \text{GS}|\hat{x}|p\rangle = \frac{2\sqrt{N_u}a\psi_n(0)}{j\pi} \qquad (F.11)$$

for odd j. $\psi_n(0)$ is the particle-hole wavefunction at $r = 0$. This is only non-zero for exited states that have even particle-hole parity (or are odd under a particle-hole transformation). This implies odd n, as for these states $\psi_n(r)$ is an even function of r. Furthermore, since ψ_n is normalized, dimensional arguments imply that $\psi_n(0) \sim \sqrt{a/\langle r\rangle}$, where $\langle r\rangle$ is the root-mean-square particle-hole separation (or the spread of the wavefunction).

Thus,

$$\langle \text{GS}|\hat{x}|p\rangle \sim a\sqrt{L/\langle r\rangle}, \qquad (F.12)$$

for odd n and odd j, and

$$\langle \text{GS}|\hat{x}|p\rangle = 0, \qquad (F.13)$$

otherwise. L is the length of the polymer.

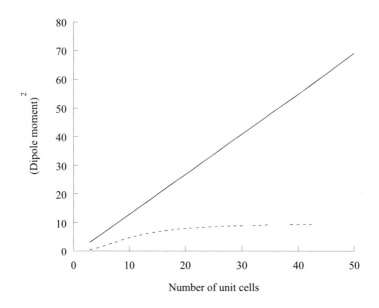

FIG. F.1. The DMRG calculated transition dipole moment squared (in arbitrary units) versus the number of unit cells for the Pariser-Parr-Pople model. $U = 10$ eV, $t = 2.5$ eV and $\delta = 0.2$. $\langle 1^1 B_u^- | \hat{x} | 1^1 A_g^+ \rangle$ (bold curve) and $\langle 2^1 A_g^+ | \hat{x} | 1^1 B_u^- \rangle$ (dashed curve). For long chains this result agrees with the weak-coupling theory. However, for chain lengths shorter than the particle-hole separations of the $|1^1 B_u^- \rangle$ and $|2^1 A_g^+ \rangle$ states the matrix element $\langle 2^1 A_g^+ | \hat{x} | 1^1 B_u^- \rangle$ shows a chain length dependence.

The square of the transition dipole moment is plotted in Fig. F.1 for an exact calculation of the Pariser-Parr-Pople model. Equation (F.13) and Fig. F.1 both show that the square of the transition dipole moment scales as L, as it must do for the oscillator strength (defined in eqn (8.6)) to satisfy the sum rule, eqn (8.7).

F.1.2   *Transitions between excited states*

The terms in the dipole operator that connect excited states are now $\hat{N}_\ell^v$ and $\hat{N}_\ell^c$. So,

$$\hat{x} = \sum_\ell x_\ell \left((N_\ell^v - 2) + N_\ell^c \right). \tag{F.14}$$

Now since,

$$\hat{x} | R + r/2, R - r/2 \rangle = r | R + r/2, R - r/2 \rangle, \tag{F.15}$$

we have that,

$$\langle p'_{n'j'} | \hat{x} | p_{nj} \rangle = \frac{1}{2a} \int dr \psi_n(r) r \psi_{n'}(r). \tag{F.16}$$

This is independent of chain length, as shown in Fig. F.1 for large chains. Notice that this matrix element connects states with the same value of j and opposite particle-hole parity. Thus, these transition dipole moments are only nonzero when $j = j'$ and $|n - n'| = $ odd.

F.2   The strong-coupling limit

In the strong-coupling limit a general excited state is described by eqn (6.30) or eqn (6.31). However, now the basis state $|R + r/2, R - r/2\rangle$ corresponds to an empty site at $R - r/2$ and a doubly occupied site at $R + r/2$ in a sea of singly occupied sites, as described in Section 6.3.

F.2.1   *Transitions between the ground state and an excited state*

From its definition (e.g. eqn (8.33)) it is evident that the dipole operator does not connect the ground state to any excited state, and thus to zeroth order in t/U the transition dipole moments to the ground state are zero in this limit. However, we know from the sum rule, eqn (8.7), that this result cannot be true to all orders of t/U. In fact, assuming that the lowest exciton state, $|p\rangle$, carries all the oscillator strength, we have,

$$\langle GS|\hat{x}|p\rangle^2 = \frac{N_e \hbar^2}{2mE_f},   \tag{F.17}$$

where $E_f \sim U$ is the transition energy. Thus,[58]

$$\langle GS|\hat{x}|p\rangle^2 = aL\frac{t}{U}.   \tag{F.18}$$

F.2.2   *Transitions between excited states*

The transition dipole moments between excited states are given by,

$$\langle p'_{n'j'}|\hat{x}|p_{nj}\rangle = \frac{1}{a}\int dr\psi_n(r)r\psi_{n'}(r),   \tag{F.19}$$

where now ψ_n are the particle-hole wavefunctions for the Mott-Hubbard excitons. The same selection rules apply as in the weak-coupling limit.

---

[58]This result can also be derived by noting that Mott-Hubbard excitons include some ground state (or covalent) character, with amplitudes $O(\sqrt{t/U})$.

APPENDIX G

VALENCE-BOND DESCRIPTION OF BENZENE

In Section 11.2.1 the electronic spectrum of benzene was discussed from the molecular orbital (or noninteracting) limit. However, as experiments and the exact solution of the Pariser-Parr-Pole model indicate, the molecular orbital approach fails to qualitatively predict the low-lying singlet spectrum. The covalent $j = 3$ transition, namely the $1^1B_{2u}^+$ state, lies energetically well below the ionic $j = 3$ and $j = 1$ transitions, namely the $1^1B_{1u}^-$ and $1^1E_{1u}^-$ states, respectively. In contrast, the molecular orbital solution predicts that these states are degenerate.

The molecular orbital approach is valid in the weak-coupling limit. In the other limit of strong-coupling the valence bond method is a more suitable approach. The basis states employed by the valence bond method are real-space states. As described in Section 5.5, at half-filling the basis states can be characterized by the number of doubly occupied sites (with the same number of empty sites). Basis states with no doubly occupied sites are classed as 'covalent', whereas basis states with one or more doubly occupied site are classed as 'ionic'. In the limit of strong electronic interactions the ionic basis states are much higher in energy than the covalent states. Thus in the strong-coupling limit we need only consider the covalent states.

The effective low-energy Hamiltonian for the purely covalent basis is the Heisenberg antiferromagnet,

$$H = J \sum_i \mathbf{S}_i \cdot \mathbf{S}_{i+1}, \tag{G.1}$$

where

$$J = \frac{4t^2}{U - V_1}. \tag{G.2}$$

For the benzene molecule with six π-orbitals the $S = 0$ subspace is spanned by five basis states. A particular nonorthogonal representation of these basis states is illustrated in Fig. G.1. There are two equivalent Kekulé structures and three equivalent Dewar structures (Coulson 1961).

Solving eqn (G.1) within this subspace we obtain the singlet spectrum listed in Table G.1. Notice that the $j = \pm 1$ (or $k = \pm \pi/3a$) excitation (corresponding to E_{1u} symmetry) is absent. The $j = 3$ (or $k = \pi/a$) state is the lowest-lying singlet excitation. Since this has same symmetry as the $1^1B_{2u}^+$ state we see that the valence bond method qualitatively predicts the lowest lying singlet excitation of benzene. This state is represented in Fig. G.1(b). Its excitation energy relative to the ground state is

(a)

$|\psi\rangle = C_1$

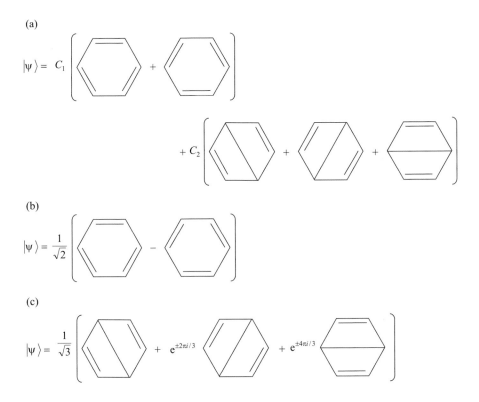

$+ C_2$

(b)

$|\psi\rangle = \dfrac{1}{\sqrt{2}}$

(c)

$|\psi\rangle = \dfrac{1}{\sqrt{3}}$

FIG. G.1. The singlet eigenstates of benzene within the valence bond covalent sub-space. Ignoring the lines representing the benzene skeleton, a line between two vertices indicates a singlet bond. The left-hand diagram in (a) shows the two equivalent Kekulé structures, while the right-hand diagram shows the three equivalent Dewar structures. (a) The $j = 0$ ($k = 0$) states with energies $(\mp\sqrt{13}/2 - 1)J$ and coefficients $C_2/C_1 = (1 \mp \sqrt{13})/6$. (b) The $j = 3$ ($k = \pi/a$) state with an energy $-3J/2$. (c) The doubly degenerate $j = \pm 2$ ($k = \pm 2\pi/3a$) states with an energy $-J/2$.

$$\frac{\sqrt{13} - 1}{2}J. \tag{G.3}$$

We emphasize that although retaining just the covalent diagrams of the valence bond method provides useful insight into the lowest-lying singlet excitation of benzene, this approach widely overestimates the excitation energies. For example, using eqns (G.2) and (G.3) with the Pariser-Parr-Pople parameters of $U = 10.06$ eV, $V_1 = 7.19$ eV, and $t = 2.539$ eV, implies that $J = 8.98$ eV and thus the excitation energy of the $1^1B_{2u}^+$ state is 11.7 eV. Since this prediction is much higher than the experimental value it indicates that the $1^1B_{2u}^+$ state must also contains some ionic character. (See Bondeson and Soos (1979) for the full

Table G.1 *The singlet spectrum of benzene using the covalent diagrams of the valence bond method*

j	k	Energy	Diagram from Fig. G.1
0	0	$-(1 + \sqrt{13}/2)J$	(a) with $C_2/C_1 = (\sqrt{13} - 1)/6$
3	π/a	$-3J/2$	(b)
± 2	$\pm 2\pi/3a$	$-J/2$	(c)
0	0	$(\sqrt{13}/2 - 1)J$	(a) with $C_2/C_1 = (\sqrt{13} + 1)/6$

Table G.2 *The triplet spectrum of benzene using the covalent diagrams of the valence bond method*

j	k	Energy
0	0	$-2.118J$
± 2	$\pm 2\pi/3a$	$-1.281J$
3	π/a	$-J$
0	0	$0.118J$
± 2	$\pm 2\pi/3a$	$0.781J$
± 1	$\pm \pi/3a$	J

valence bond analysis of the singlet spectrum.)

The covalent valence bond method also fails to qualitatively predict the triplet spectrum, as it predicts the lowest-lying triplet to be at $k = 0$, whereas both experiment and the exact solution of the Pariser-Parr-Pople model place it at $k = \pi/a$. The covalent valence bond predictions are listed in Table G.2.

APPENDIX H

DENSITY MATRIX RENORMALIZATION GROUP METHOD

The density matrix renormalization group (DMRG) method is an efficient and accurate Hilbert space truncation procedure (White 1992; 1993) that can be used to solve quantum mechanical models on very large systems. It is particularly suited for one-dimensional quantum lattice models, such as the π-electron models discussed in this book. This appendix contains a brief review of the DMRG method relevant for these models. A full discussion of the method and its various applications may be found in (Peschel *et al.* 1999), (Dukelsky and Pittel 2004), or (Schollwöck 2005).

H.1 Introduction to the real-space method

H.1.1 *Infinite algorithm method*

Consider a linear chain of N sites composed of four *blocks*, labelled $i = 1, \ldots, 4$. At present we consider symmetric systems with open boundary conditions so that blocks 1 and 4, and blocks 2 and 3 are equivalent. This is illustrated in Fig. H.1. A general state of the block i is denoted by $|m_i\rangle^{(i)}$, where m_i is a shorthand label for the quantum numbers of the block (e.g. the conserved quantum numbers, spin and charge, and any other state index). Block i has N_i sites and its Hilbert space is spanned by M_i states.

Block 1 is augmented by block 2 to form the *system block* with $N_s = N_1 + N_2$ sites and a Hilbert space of $M_s = M_1 M_2$ states. Similarly, block 4 is augmented by block 3 to form the *environment block* with $N_e = N_3 + N_4$ sites and a Hilbert space of $M_e = M_3 M_4$ states.

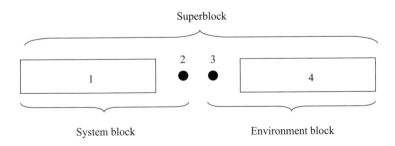

FIG. H.1. The component blocks for the infinite algorithm method of the DMRG technique for open boundary conditions.

If this procedure is repeated so that the left- and right- hand blocks are grown by sequentially augmenting them with the middle blocks, the Hilbert space size of these blocks would grow exponentially in size as a function of their physical size. The goal is therefore to truncate the system block Hilbert space so that $M_s \to \tilde{M}_s \approx M_1$ at each iteration. This goal is achieved by the DMRG method.

A general system block state, $|m_s\rangle^{(s)}$, is a direct product of states of blocks 1 and 2,

$$|m_s\rangle^{(s)} = |m_1\rangle^{(1)}|m_2\rangle^{(2)}. \tag{H.1}$$

Similarly, a general environment block state, $|m_e\rangle^{(e)}$, is a direct product of states of blocks 3 and 4,

$$|m_e\rangle^{(e)} = |m_3\rangle^{(3)}|m_4\rangle^{(4)}. \tag{H.2}$$

Together, the system and environment blocks constitute the *superblock*. The total Hamiltonian is applied to the superblock. Thus, unlike the constituent blocks, the superblock is specified by conserved quantum numbers (e.g. spin and charge), and consequently a subbasis of the entire Hilbert space is constructed from a subbasis of the direct product of the system and environment blocks. Denoting a subbasis of the superblock as $\{|m\rangle\}$ we have

$$\{|m\rangle\} = \{|m_s\rangle^{(s)}|m_e\rangle^{(e)}\}. \tag{H.3}$$

We may thus express a superblock eigenstate as,

$$
\begin{aligned}
|\Psi\rangle &= \sum_m \Psi(m)|m\rangle \\
&= \sum_{m_s}^{M_s} \sum_{m_e}^{M_e} \Psi(m_s, m_e)|m_s\rangle^{(s)}|m_e\rangle^{(e)},
\end{aligned} \tag{H.4}
$$

where $\Psi(m)$ is determined by diagonalizing the superblock Hamiltonian.

We wish to truncate and rotate the Hilbert space of the system block subject to retaining an optimal representation of $|\Psi\rangle$. Denoting the size of the truncated superblock Hilbert space as $\tilde{M}_s$ and the approximate, optimal superblock state as $|\tilde{\Psi}\rangle$ we have,

$$|\tilde{\Psi}\rangle = \sum_{m_s}^{\tilde{M}_s} \sum_{m_e}^{M_e} \tilde{\Psi}(m_s, m_e)|m_s\rangle\rangle^{(s)}|m_e\rangle^{(e)}, \tag{H.5}$$

where $\{|m_s\rangle\rangle^{(s)}\}$ is the rotated optimal basis of the system block. By optimal we mean that $\langle\Psi|\tilde{\Psi}\rangle$ is maximized.

It can be shown that $|\tilde{\Psi}\rangle$ is optimized if the rotated and truncated basis of the system block are the $\tilde{M}_s$ eigenstates with the highest eigenvalues of the system block reduced density matrix. These states are the most 'probable' for describing the approximate state, $|\tilde{\Psi}\rangle$, in the truncated basis.

The system block reduced density matrix is,

$$\rho_{m_s m_s'} = \sum_{m_e} \Psi(m_s, m_e) \Psi(m_s', m_e), \qquad (H.6)$$

where the sum is over the environment block states.

In the next DMRG iteration the new block 1 is the old system block (and the new block 4 is its spatial reflection). Thus, the overall chain has grown by $N_2 + N_3$ sites. Furthermore, since the Hilbert space of the new blocks 1 and 4 are essentially the same as the old blocks, although their physical size has increased, the Hilbert space of the superblock remains essentially constant as it grows. Thus, it is possible to perform highly accurate calculations on chains much longer than is possible by conventional exact diagonalization techniques. Since the DMRG method is a variational technique, more accurate approximations can be obtained by increasing the number of states retained in any block.

In the following sections we briefly describe a few technical details, the finite lattice algorithm for improving the DMRG accuracy for a particular system size, and extensions of the DMRG method.

H.1.2 *Rotation and truncation of the basis*

We require the matrix elements of the Hamiltonian and other operators pertaining to the system block in the new, truncated basis. These matrix elements are denoted as

$$\tilde{O}(m_s', m_s) = \, ^{(s)}\langle\langle m_s' | \hat{O} | m_s \rangle\rangle^{(s)}. \qquad (H.7)$$

By a similarity transformation,

$$\tilde{O} = S O S^\dagger, \qquad (H.8)$$

where $S^\dagger$ is the $M_s \times \tilde{M}_s$ rectangular matrix whose $\tilde{M}_s$ columns are the density matrix eigenstates expressed in the old basis, and

$$O(m_s', m_s) = \, ^{(s)}\langle m_s' | \hat{O} | m_s \rangle^{(s)}. \qquad (H.9)$$

H.1.3 *Symmetries and excited states*

Eigenstates of the superblock Hamiltonian are conveniently calculated using a sparse matrix diagonalization routine, such as the conjugate gradient method. Excited states specified by certain conserved quantum numbers within a particular symmetry sector may be found by a Gram-Schmidt orthogonalization procedure. Relevant symmetries for conjugated polymers are spatial inversion, particle-hole symmetry and spin-flip symmetry (which distinguishes between singlet and triplet states). The construction of symmetry adapted trial states in the conjugate gradient routine is greatly facilitated by using sparse block-symmetry operators. Their construction is discussed in the next section.

H.1.3.1 *Sparsity of the block-symmetry operators* Suppose that $\hat{O}^{(s)}$ is a symmetry operator of the system block (for example, particle-hole or spin-flip symmetry). Then it can be shown that

$$\hat{O}^{(s)}\hat{\rho}_Q O^{(s)\dagger} = \hat{\rho}_{\bar{Q}}, \tag{H.10}$$

where $\hat{\rho}_Q$ is the system block reduced density operator for a particular conserved quantum number, Q, (e.g. spin and/or charge) and $\bar{Q}$ is the complementary quantum number obtained via the action of the symmetry operator, $\hat{O}^{(s)}$.

Equation (H.10) has two significant consequences. To explain these, we modify the notation for the system blocks states from $|n_s\rangle\rangle^{(s)}$ to $|Q, n_s\rangle\rangle^{(s)}$, where Q denotes the conserved quantum numbers of the block and n_s is now a supplementary index. Then

1. If $Q = \bar{Q}$ then $\hat{O}^{(s)}$ commutes with $\hat{\rho}_Q$ and the eigenstates of $\hat{\rho}_Q$ are simultaneously eigenstates of $\hat{O}^{(s)}$. Thus,

$$\hat{O}^{(s)}|Q, n_s\rangle\rangle^{(s)} = \pm|Q, n_s\rangle\rangle^{(s)}. \tag{H.11}$$

2. If $Q \neq \bar{Q}$ then all the states with complementary quantum numbers may be defined as

$$|\bar{Q}, n_s\rangle\rangle^{(s)} = \hat{O}^{(s)}|Q, n_s\rangle\rangle^{(s)}. \tag{H.12}$$

Equations (H.11) and (H.12) imply that the rotated block symmetry operator is sparse.

H.1.4   *Finite algorithm method*

Accuracy of the wavefunction and energy at a particular superblock size can be improved by performing finite lattice sweeps. That is, DMRG calculations are performed with blocks 1 and 4 of different sizes. A sweep from left to right starts with block 1 containing N_2 sites (that is, it is a copy of block 2), and block 4 containing $N - 3N_2$ sites. Block 1 is continually augmented until it has $N - 3N_2$ sites. This procedure shuffles the Hilbert spaces of the blocks, resulting in improved accuracy.

H.1.5   *Application to linear polyenes*

The application of the DMRG technique to solving the Pariser-Parr-Pople or Pariser-Parr-Pople-Peierls model for linear polyenes is relatively standard. In this case blocks 2 and 3 are single sites representing a single π-orbital. Each orbital has four allowed spin states, so the Hilbert space of the (augmented) system block is $4 \times M_1$. Using the infinite lattice method means that the chains are symmetric, so inversion symmetry, as well as particle-hole and spin-flip symmetries may be imposed to target $^1B_u^-$, $^3B_u^+$, $^1A_g^+$ states, etc. More challenging applications of the DMRG method arise when studying electron-phonon problems and phenyl-based light emitting polymers. In these cases it is necessary to perform a local Hilbert space truncation, as described in the next section.

H.2   Local Hilbert space truncation

The DMRG method outlined above is a procedure to calculate the low-lying eigenstates of large systems. However, the concept of constructing an optimal basis for a block (previously, the combined blocks 1 and 2) has wider generality (Zhang *et al.* 1998), particularly if the block is a natural repeat unit or moiety of a molecular chain. Often the full Hilbert space of block 2 will be too large to augment with block 1, and so an optimal truncation of that block is needed. Examples of these kinds of problems include:

1. Electron-phonon problems - discussed in Chapter 10 (Barford *et al.* 2002a), where the Hilbert space size of a single site is 4×(number of site-phonons+1).
2. Light emitting polymers - discussed in Chapter 11 (Bursill and Barford 2002, 2005), where the Hilbert space size of the phenyl rings in π-electron models is 2^{12}.
3. Larger basis semiemprical quantum chemistry models, such as INDO, where the Hilbert space size of the atoms or moieties is very large.

The procedure for constructing an optimally truncated Hilbert space of a single block (usually block 1 or 2) is a straightforward generalization of the DMRG method. Suppose that the Hilbert space of block j is to be truncated. Then we construct an environment block state from the remaining block states,

$$|m_e\rangle^{(e)} = \prod_{i\neq j} |m_i\rangle^{(i)}, \tag{H.13}$$

and the direct product space of the superblock is then

$$\{|m\rangle\} = \{|m_j\rangle^{(j)}|m_e\rangle^{(e)}\}. \tag{H.14}$$

The superblock eigenstate is

$$|\Psi\rangle = \sum_{m_j}^{M_j}\sum_{m_e}^{M_e} \Psi(m_j, m_e)|m_j\rangle^{(j)}|m_e\rangle^{(e)}. \tag{H.15}$$

Again the optimal basis for the single block j are the eigenstates with the highest eigenvalues of its reduced density matrix, defined by,

$$\rho_{m_j m_j'} = \sum_{m_e} \Psi(m_j, m_e)\Psi(m_j', m_e). \tag{H.16}$$

REFERENCES

Abe, A. (1993) *Relaxation in Polymers*, edited by Kobayashi, T. World Scientific, Singapore.

Abe, A., Yu, J., and Su, W-P. (1992) *Phys. Rev. B* **45**, 8264.

Affleck, A. (1997) *Dynamical Properties of Unconventional Magnetic Systems*, NATO ASI, Geilo, Norway.

Ambrosch-Draxl, C., Majewski, J. A., Vogl, P., and Leising, G. (1995) *Phys. Rev. B* **51**, 9668.

Anderson, P. W. (1984) *Basic Notions of Condensed Matter Physics*, Benjamin-Cummings, Menlo Park.

Artacho, E., Rohfling, M., Côte, M., Haynes, P. D., Needs, R. J., and Molteni, C. (2004) *Phys. Rev. Lett.* **93**, 116401.

Atkins, P. W. and Friedman, R. S. (1997) *Molecular Quantum Mechanics*, 3rd edition, Oxford University Press, Oxford.

Baeriswyl, D. (1985) *Theoretical Aspects of Band Structures and Electronic Properties of Pseudo-One-Dimensional Solids*, edited by Kamimura, H. R. Reidel Publishing Company, Dordrecht.

Baeriswyl, D. and Maki, K. (1985) *Phys. Rev. B* **31**, 6633.

Baeriswyl, D., Campbell, D. K., and Mazumdar, S. (1992) *Conjugated Conducting Polymers*, edited by Kiess, H. G., Springer-Verlag, Berlin.

Ball, R., Su, W-P., and Schrieffer, J. R. (1983) *J. de Phys.* **44**, C3-429.

Barford, W. (2002) *Phys. Rev. B* **65**, 205118.

Barford, W. (2004) *Phys. Rev. B* **70**, 205204.

Barford, W., Bursill, R. J., and Lavrentiev, M. Yu (1998) *J. Phys.: Condens. Matt.* **10**, 6429.

Barford, W., Bursill, R. J., and Lavrentiev, M. Yu. (2001) *Phys. Rev. B* **63**, 195108.

Barford, W., Bursill, R. J., and Lavrentiev, M. Yu (2002a) *Phys. Rev. B* **65**, 75107.

Barford, W., Bursill, R. J., and Smith, R. W. (2002b) *Phys. Rev. B* **66**, 115205.

Barford, W., Bursill, R. J., and Yaron, D. (2004) *Phys. Rev. B* **69**, 155203.

Barth, S., Bässler, H., Scherf, U., and Müllen, K. (1998) *Chem. Phys. Lett.* **288**, 147.

Bässler, H. (2000) *Semiconducting Polymers: Chemistry, Physics and Engineering*, edited by van Hutten, G., Wiley-VCH, Berlin.

Beljonne, D., Shuai, Z., Friend, R. H., and Brédas, J. L. (1995) *J. Chem. Phys.* **102**, 2042.

Beljonne, D., Shuai, Z., Cornil, J., dos Santos, D. A., and Brédas, J. L. (1999) *J. Chem. Phys.* **111**, 2829.

Beljonne, D., Ye, D. A., Shuai, Z., and Brédas, J. L. (2004) *Adv. Funct. Mater.* **14**, 684.

Birks, J. B. (1970) *Photophysics of Aromatic Molecules*, Wiley, London.

Bishop, A. R., Campbell, D. K., and Fesser, K. (1981) *Mol. Cryst. Liq. Cryst.* **77**, 252.

Bishop, A. R., Campbell, D. K., Lomdahl, P., Horovitz, B., and Phillpot, S. R. (1984) *Phys. Rev. Lett.* **52**, 671.

Bondeson, S. R. and Soos, Z. G. (1979) *J. Chem. Phys.* **71**, 3807.

Brazovoskii, S. A. and Kirova, N. N. (1981) *JETP Lett.* **33**, 4.

Burin, A. L. and Ratner, M. A. (1998) *J. Chem. Phys.* **109**, 6092.

Burroughes, J. H., Bradley, D. D. C., Brown, A. R., Marks, R. N., Mackay, K., Friend, R. H., Burn, P. L., and Holmes, A. D. (1990) *Nature* **347**, 539.

Bursill, R. J. and Barford, W. (1999) *Phys. Rev. Lett.* **82**, 1514.

Bursill, R. J. and Barford, W. (2002) *Phys. Rev. B* **66**, 205112.

Bursill, R. J. and Barford, W. (2005) unpublished.

Bursill, R. J., Castleton, C., and Barford, W. (1998) *Chem. Phys. Lett.* **294**, 305.

Butcher, P. N. and Cotter, D. (1990) *The Elements of Nonlinear Optics*, Cambridge University Press, Cambridge.

Cadby, A. J. and Martin, S. (2004) unpublished.

Cadby, A. J., Lane, P. A., Mellor, H., Martin, S. J., Grell, M. Giebler, C., Bradley, D. D. C., Wohlgenannt, M., An, C., and Vardeny, Z. V. (2000) *Phys. Rev. B* **62**, 15604.

Campbell, D. K. and Bishop, A. R. (1981) *Phys. Rev. B* **24**, 4859.

Cao, Y., Parker, I. D., Gang, Y., Zhang, C., and Heeger, A. J. (1999) *Nature* **397**, 414.

Castelton, C. and Barford, W. (1999) *Synth. Met.* **101**, 520.

Castleton, C. W. M. and Barford, W. (2002) *J. Chem. Phys.* **117**, 3570.

Chandross, M. and Mazumdar, S. (1997) *Phys. Rev. B* **55**, 1497.

Chandross, M., Mazumdar, S., Jeglinski, S., Wei, X., Vardeny, Z. V., Kwock, E. W., and Miller, T. M. (1994) *Phys. Rev. B* **50**, 14702.

Chandross, M., Mazumdar, S., Liess, M., Lane, P. A., Vardeny, Z. V., Hamaguchi, M., and Yoshino, K. (1997) *Phys. Rev. B* **55**, 1486.

Chiang, C. K., Fincher, C. R. Jr., Park, Y. W., Heeger, A. J., Shirakawa, H., and Louis, E. J. (1977) *Phys. Rev. Lett.* **39**, 1098.

Cohen-Tannoudji, C., Diu B., and Laloë, F. (1977) *Quantum Mechanics*, Hermann, Paris.

Coll, C. F. (1974) *Phys. Rev. B* **9**, 2150.

Coulson, C. A. (1961) *Valence*, Oxford University Press, Oxford.

Coulson, C. A., and Dixon, W. T. (1961) *Tetrahedron* **17**, 215.

Dexter, D. L. (1953) *J. Chem. Phys.* **21**, 836.

Dhoot, A. S., Ginger, D. S., Beljonne, D., Shuai, Z., and Greenham, N. C. (2002) *Chem. Phys. Lett.* **360**, 195.

DiBartolo, B. (1980) *Radiationless Processes*, Plenum Press, New York.

Dixit, S. N. and Mazumdar, S. (1984) *Phys. Rev. B* **29**, 1824.

Dixit, S. N., Guo, D., and Mazumdar, S. (1991) *Phys. Rev. B* **43**, 6781.

Dukelsky, J. and Pittel, S. (2004) *Rep. Prog. Phys.* **67**, 513.

Ehrenfreund, E., Vardeny, Z., Brafman, O., and Horovitz, B. (1987) *Phys. Rev. B* **18**, 5756.

Emin, D. and Holstein, T. (1976) *Phys. Rev. Lett.* **36**, 323.

Essler, F. H. L., Gebhard, F., and Jeckelmann, J. (2001) *Phys. Rev. B* **64**, 125119.

Essler, F. H. L., Frahm, H., Göhmann, F., Klümper, A., and Korepin, V. E. (2005) *The One-Dimensional Hubbard Model*, Cambridge University Press, Cambridge.

Fann, W-S., Benson, S., Madey, J. M. J., Etemad, S., Baker, G. L., and Kajzar, F. (1989) *Phys. Rev. Lett.* **62**, 1492.

Förster, T. (1951) *Fluoreszenz Organischer Verbindungen*, van den Hoek and Rupprecht, Gottingen.

Fradkin, E. and Hirsch, J. E. (1983) *Phys. Rev. B* **27**, 1680.

Friend, R. H., Bradley, D. D. C., and Townsend, P. D. (1987) *J. Phys. D: Appl. Phys.* **20**, 1367.

Fröhlich, H. (1954) *Proc. Roy. Soc.* London **A223**, 296.

Frolov, S., Liess, M., Lane, P., Gellermann, W., and Vardeny, Z. (1997) *Phys. Rev. Lett.* **78**, 4285.

Frolov, S., Bao, Z., Wohlgenannt, M., and Vardeny, Z. V. (2002) *Phys. Rev. B* **65**, 205209.

Fulde, P. (1993) *Electron Correlations in Molecules and Solids*, Springer-Verlag, Berlin.

Gallagher, F. B. and Mazumdar, S. (1997) *Phys. Rev. B* **56**, 15025.

Gartstein, Yu. N., Rice, M. J., and Conwell, E. M. (1995) *Phys. Rev. B* **52**, 1683.

Gebhard, F., Bott, K., Scheidler, M., Thomas, P., and Koch, S. W. (1997) *Philos. Mag. B* **75**, 13.

Giamarchi, T. (2003) *Quantum Physics in One Dimension*, Oxford University Press, Oxford.

Gordon, M. and Ware, W. R. (1975) eds. *The Exciplex*, Academic Press Inc., New York.

Grabowski, M., Hone, D., and Schrieffer, J. R. (1985) *Phys. Rev. B* **31**, 7850.

Halvorson, C. and Heeger, A. J. (1993) *Chem. Phys. Lett.* **216**, 488.

Harris, A. B. and Lange, R. V. (1967) *Phys. Rev.* **157**, 295.

Hayden, G. W. and Mele, E. J. (1986) *Phys. Rev. B* **34**, 5484.

Heeger, A. J., Kivelson, S., Schrieffer, J. R., and Su, W-P. (1988) *Rev. Mod. Phys.*, **60**, 781.

Heeger, A. J. (2000) http://www.nobel.se/chemistry/laureates/2000/heeger-lecture.pdf.

Heflin, J. R., Wong, K. Y., Zamani-Khamiri, O., and Garito, A. F. (1988) *Phys. Rev. B* **38**, 1573.

Henderson, B. and Imbusch, G. F. (1989) *Optical Spectroscopy of Inorganic Solids*, Oxford University Press, Oxford.

Hertel, D., Setayesh, S., Nothofer, H.-G., Scherf, U., Müllen, K., and Bässler, H. (2001) *Advanced Materials*, **13**, 65.

Hirsch, J. E. (1983) *Phys. Rev. Lett.* **51**, 296.

Ho, P. K. H., Kim, J-S., Burroughes, J. H., Becker, H., Sam, F. Y. L., Brown, T. M., Cacialli, F., and Friend, F. H. (2000) *Nature* **404**, 481.

Holstein, T. (1959) *Ann. Phys.* (N.Y.) **8**, 343.

Hong, T-M. and Meng, H-F. (2001) *Phys. Rev. B* **63**, 075206.

Horsch, P. (1981) *Phys. Rev. B* **24**, 7351.

Huang, K. and Rhys, A. (1950) *Proc. Roy. Soc.* A**204** , 406.

Hückel. E. (1931) *Z. Phys.* **70**, 204.

Hückel. E. (1932) *Z. Phys.* **76**, 623.

Hudson, B. S. and Kohler, B. E. (1972) *Chem. Phys. Lett.* **14**, 299.

Itô, T., Shirakawa, H., and Ikeda, S. (1974) *J. Polym. Sc., Polym. Chem. Ed.* **12**, 11.

Kadashchuk, A., Vakhnin, A., Blonski, I., Beljonne, D., Shuai, Z., Brédas, J. L., Arkhipov, V. I., Heremans, P., Emelianova, E. V., and Bässler, H. (2004) *Phys. Rev. Lett.* **93**, 066803.

Kahlert, H., Leitner, O., and Leising, G. (1987) *Synth. Met.* **17**, 467.

Karabunarliev, S. and Bittner, E. R. (2003) *J. Chem. Phys.* **119**, 3988.

Keil, T. (1965) *Phys. Rev.* **140**, A601.

Kirova, N. and Brazovskii, S. (2004) *Synth. Mets.* **141**, 139.

Kirova, N., Brazovskii, S., and Bishop, A. R. (1999) *Synth. Mets.* **100**, 29.

Knox, R. S. (1963) *Solid State Physics*, Academic, New York, Suppl. 5.

Kohler, B. E. (1988) *J. Chem. Phys.* **88**, 2788.

Köhler, A. and Wilson, J. (2003) *Organic Electronics* **4**, 179.

Köhler, A. and Beljonne, D. (2004) *Advanced Functional Materials* **14**, 11.

Konig, G. and Stollhoff, G. (1990) *Phys. Rev. Lett.* **65**, 1239.

Landau, L. D. and Lifshitz, E. M. (1977) *Quantum Mechanics*, 3rd edition, Pergamon Press, Oxford.

Lane, P. A., Liess, M., Vardeny, Z. V., Hamaguchi, M., Ozaki, M., and Yoshino, K. (1997) *Synth. Met.* **84**, 641.

Lennard-Jones, J. E. (1937) *Proc. Roy. Soc. A.* **158**, 280.

Liess, M., Jeglinski, S., Vardeny, Z. V., Ozaki, M., Yoshino, K., Ding, Y., and Barton, T. (1997) *Phys. Rev. B* **56**, 15712.

Longuet-Higgins, H. C. and Salem, L. (1959) *Proc. Roy. Soc.* London **A251**, 172.

Lorentzon, J., Malmqvist, P., Fülscher, M., and Roos, B. O. (1995) *Theor. Chim. Acta.* **91**, 91.

Loudon, R. (1959) *Amer. J. Phys.* **27**, 649.

Loudon, R. (2000) *The Quantum Theory of Light*, 3rd edition, Oxford University Press, Oxford.

MacDiarmid, A. G. (2000) http://www.nobel.se/chemistry/laureates/2000/macdiarmid-lecture.pdf.

Markus, R. A. (1964) *Ann. Rev. Phys. Chem.* **15** 155.

Martin, S. J., Bradley, D. D. C., Lane, P. A., Mellor, H., and Burn, P. L. (1999) *Phys. Rev. B* **59**, 15133.

Mathy, A., Ueberhofen, K., Schenk, R., Gregorius, H., Garay, R., Müllen, K., and Bubeck, C. (1996) *Phys. Rev. B* **53**, 4367.

May, V. and Kühn, O. (2000) *Charge and Energy Dynamics in Molecular Systems*, Wiley-VCH, Berlin.

Mazumdar, S. and Campbell, D. K. (1985) *Phys. Rev. Lett.* **55**, 2067.

Mazumdar, S. and Chandross, M. (1997) *Primary Photoexcitations in Conjugated Polymers: Molecular Exciton versus Semiconductor Band Model*, edited by Sariciftci, N. S., World Scientific, Singapore.

Mazumdar, S. and Dixit, S. N. (1983) *Phys. Rev. Lett.* **51**, 292.

Mazumdar, S. and Soos, Z. G. (1979) *Synth. Met.* **1**, 77.

Meskers, S. C. J., Hübner, J., Oestreich, M., and Bässler, H. (2001) *J. Phys. Chem. B* **105**, 9139.

McKenzie, R. H. and Wilkin, J. W. (1992) *Phys. Rev. Lett.* **69**, 1085.

Miller, E. K., Yoshida, D., Yang, C. Y., and Heeger, A. J. (1999a) *Phys. Rev. B* **59**, 4661.

Miller, E. K., Maskel, G. S., Yang, C. Y., and Heeger, A. J. (1999b) *Phys. Rev. B* **60**, 8028.

Misurkin, I. A. and Ovchinnikov, A. A. (1971) *Sov. Phys.-Solid State* **12**, 2031.

Monkman, A. P., Burrows, H. D., Hartwell, L. J., Horsburgh, L. E., Hamblett, I., and Navaratnam, S. (2001) *Phys. Rev. Lett.* **86**, 1358.

Moore, E. and Yaron, D., (1998) *J. Chem. Phys.* **109**, 6147.

Moore, E., Barford, W., and Bursill, R. J. (2005) *Phys. Rev. B* **71**, 115107.

Movaghar, B., Grünewald, M., Ries, B., Bässler, H., and Würtz, D. (1986) *Phys. Rev. B* **33**, 5545.

Mukamel, S. (1995) *Principles of Nonlinear Optical Spectroscopy*, Oxford University Press, New York.

Mukhopadhyay, D., Hayden, G. W., and Soos, Z. G., (1995) *Phys. Rev. B* **51**, 9476.

Nakano, T., and Fukuyama, H. (1980) *J. Phys. Soc. Japan* **49**, 1679.

Niko, A., Zojer, E., Meghdadi, F., Ambrosch-Draxl, C., and Leising, G. (1999) *Synth. Met.* **101**, 662.

Ooshika, Y. (1957) *J. Phys. Soc. Japan* **12**, 1246.

Ooshika, Y. (1959) *J. Phys. Soc. Japan* **14**, 747.

Ovchinnikov, A. A., Ukrainskii, I. I., and Kventsel, G. V. (1973) *Sov. Phys. Usp.* **15**, 575.

Pariser, R. and Parr, R. G. (1953a) *J. Chem. Phys.* **21**, 466.

Pariser, R. and Parr, R. G. (1953b) *J. Chem. Phys.* **21**, 767.

Peierls, R. E. (1955) *Quantum Theory of Solids*, Oxford University Press, Oxford.

Peschel, I., Wang, X., Kaulke, M., and Hallberg, K. (1999) editors *Density Matrix Renormalization*, Springer, Berlin.

Phillpot, S. R. , Bishop, A. R., and Horovitz, B. (1989) *Phys. Rev. B* **40**, 1839.

Pope, M. and Swenberg, C. E. (1999) *Electronic Processes in Organic Crystals and Polymers*, 2nd edition, Oxford University Press, New York.

Pople, J. A. (1953) *Trans. Farad. Soc.* **49**, 1375.

Pople, J. A. (1954) *Proc. Phys. Soc.* **68**, 81.

Pople, J. A. and Walmsley, S. H. (1962) *Molec. Phys.* **5**, 15.

Race, A., Barford, W., and Bursill, R. J. (2001) *Phys. Rev. B* **64**, 035208.

Race, A., Barford, W., and Bursill, R. J. (2003) *Phys. Rev. B* **67**, 245202.

Ramasesha, S. and Soos, Z. G. (1984) *J. Chem. Phys.* **80**, 3278.

Rehwald, W. and Kiess H. G. (1992) *Conjugated Conducting Polymers*, edited by Kiess, H. G., Springer-Verlag, Berlin.

Rice, M. J. and Gartstein, Yu. N. (1994). *Phys. Rev. Lett.* **73**, 2504.

Rohlfing, M. and Louie, S. G. (1999) *Phys. Rev. Lett.* **82**, 1959.

Salem, L. (1966) *The Molecular Orbital Theory of Conjugated Systems*, N. A. Benjamin Inc., New York.

Schollwöck, U. (2005) *Rev. Mod. Phys.* (in press).

Schulten, K. and Karplus, M. (1972) *Chem. Phys. Lett.* **14**, 299.

Sebastian, L. and Weiser, G. (1981) *Phys. Rev. Lett.* **46**, 1156.

Segal, M., Baldo, M. A., Holmes, R. J., Forrest, S. R., and Soos, Z. G. (2003) *Phys. Rev. B* **68**, 075211.

Shirakawa, H. (2000) http://www.nobel.se/chemistry/laureates/2000/shirakawa-lecture.pdf.

Shuai, Z., Bredas, J-L., Pati, S. K., and Ramesesha, S. (1997) *Phys. Rev. B* **56**, 9298.

Shukla, A., Ghosh, H., and Mazumdar, S. (2003) *Phys. Rev. B* **67**, 245203.

Song, K. S. and Williams, R. T. (1993) *Self-Trapped Excitons*, Springer, Berlin.

Soos, Z. G. and Ramasesha, S. (1989) *J. Chem. Phys.* **90**, 1067.

Soos, Z. G., Ramasesha, S., and Galvão, D. S. (1993) *Phys. Rev. Lett.* **71**, 1609.

Su, W-P. (1982) *Solid State Comm.* **42**, 497.

Su, W-P. (1995) *Phys. Rev. Lett.* **74**, 1167.

Su, W-P. and Schrieffer, J. R. (1980) *Proc. Natl. Acad. Sci. U.S.A.*, **77**, 5626.

Su, W-P., Schrieffer, J. R., and Heeger, A. J. (1979) *Phys. Rev. Lett.* **42**, 1698.

Su, W-P., Schrieffer, J. R., and Heeger, A. J. (1980) *Phys. Rev. B* **22**, 2099.

Surján, P. R. (1989) *Second Quantized Approach in Quantum Chemistry*, Springer-Verlag, Berlin.

Szabo, A. and Ostlund, N. S. (1996) *Modern Quantum Chemistry*, Dover, Mineola.

Takimoto, J. and Sasai, M. (1989) *Phys. Rev. B* **39**, 8511.

Takayama, H., Yin-Lui, Y. R., and Maki, K. (1980) *Phys. Rev. B* **21**, 2388.

Tandon, K., Ramasesha, S., and Mazumdar, S. (2003) *Phys. Rev. B* **67**, 045109.

Tavan, P. and Schulten, K. (1987) *Phys. Rev. B* **36**, 4337.

Tinkham, M. (1964) *Group Theory and Quantum Mechanics*, McGraw Hill, New York.

Tretiak, S., Saxena, A., Martin, R. L., and Bishop, A. R. (2002) *Phys. Rev. Lett.*, **89**, 097402.

Tretiak, S., Saxena, A., Martin, R. L., and Bishop, A. R. (2003) *Proc. Natl. Acad. Sci. U.S.A.* **100**, 2185.

Tsvelik, A. M. (1995) *Quantum Field Theory in Condensed Matter Physics*, Cambridge University Press, Cambridge.

Toyazawa, Y. and Shinozuka, Y. (1980) *J. Phys. Soc. Japan* **48**, 472.

Ukrainskii, I. I. (1978) *Sov. Phys.-Theo. Maths. Phys.* **32**, 816.

van der Horst, J-W., Bobbert, P. A., de Long, P. H. L., Michels, M. A. J., Brocks, G., and Kelly, P. J. (2000) *Phys. Rev. B* **61**, 15817.

van der Horst, J-W., Bobbert, P. A., Michels, M. A. J., and Bässler, H. (2001) *J. Chem. Phys.* **114**, 6950.

Vardeny, Z. V. (1993) *Relaxation in Polymers*, edited by Kobayashi, T., World Scientific, Singapore.

Walker, A.B., Kambili, A., and Martin, S. J. (2002) *J. Phys.: Condens. Matter* **14** 9825.

Weibel, J. D. and Yaron, D. (2002) *J. Chem. Phys.* **116**, 6846.

Weiser, G. and Horváth, Á. (1997) *Primary Photoexcitations in Conjugated Polymers: Molecular Exciton versus Semiconductor Band Model*, edited by Sariciftci, N. S., World Scientific, Singapore.

Wen, G-Z. and Su, W-P. (1997) *Relaxations of Excited States and Photo-induced Structural Phase Transitions*, edited by Nasu, K., Springer-Verlag, Berlin.

White, S. R. (1992) *Phys. Rev. Lett.* **69**, 2863.

White, S. R. (1993) *Phys. Rev. B* **48**, 10 345.

Wilson, J. S., Dhoot, A. S., Seeley, A. J. A. B., Khan, M. S., Köhler, A., and Friend, R. H. (2001) *Nature* 413, 828.

Wohlgenannt, M., and Vardeny, Z. V. (2003) *J. Phys.: Condens. Matter* **15**, R83.

Wohlgenannt, M., Jiang, X. M., Vardeny, Z. V., and Janssen, R. A. J. (2002) *Phys. Rev. B* **88**, 197401.

Wohlgenannt, M., Jiang, X. M., and Vardeny, Z. V. (2004) *Phys. Rev. B* **69**, 241204.

Xu, S., Klimov, V. I., Kraabel, B., Wang, H., and McBranch, D. W. (2001) *Phys. Rev. B* **64**, 193201.

Ye, A., Shuai, Z., and Brédas, J. L. (2002) *Phys. Rev. B* **65**, 045208.

Zhang, C., Jeckelmann, E., and White, S. R. (1998) *Phys. Rev. Lett.* **80**, 2661.

Zheng, W., Hamer, C. J., Singh, R. R. P., Trebst, S., and Monien, H. (2001) *Phys. Rev. B* **63**, 144411.

Ziman, J. M. (1972) *Principles of the Theory of Solids*, Cambridge University Press, Cambridge.

Zojer, E., Cornil, J., Leising, G. Fink, J., and Brédas, J. L., (1999) *Phys. Rev. B* **59**, 7957.

Zojer, E., Koch, N., Puschnig, P., Meghdadi, F., Niko, A., Resel, R., Ambrosch-Draxl, C., Knupfer, M., Fink, J., Brédas, J. L., and Leising, G. (2000) *Phys. Rev. B* **61**, 16538.

INDEX